W9-CMR-722

# of Virginia

Keith Frye
Illustrated by Ramesh Venkatakrishnan

MOUNTAIN PRESS PUBLISHING COMPANY
MISSOULA 1986

ROADSIDE GEOLOGY SERIES
Editorial Directors:
David Alt and Donald Hyndman

Library of Congress Cataloging-in-Publication Data

Frye, Keith, 1935—
Roadside geology of Virginia.

Bibliography: p.
Includes index.
1. Geology—Virginia. I. Title.
QE173.F78 1986 557.55 86-8755
ISBN 0-87842-199-8 (pbk.)

1 3 5 7 9 8 6 4 2

FOREWORD

The great natural beauty of Virginia, from the Atlantic Coast on the east to the Appalachian Plateaus on the west, is the result of geologic processes that have formed this part of the Earth for more than one billion years. Plains, plateaus, and mountains have been formed by the weathering and erosion of a diverse geologic terrane, with the tougher, more resistant rocks standing high above those that are softer and weaker. Great rivers wend their way through Virginia, cutting spectacular gorges through the mountains in search of the sea. Man, too, in constructing major roadways has cut into the Earth like the streams and rivers, thereby revealing the geology of the uppermost part of the crust.

Although all of us are at least passingly aware of the beauty of the Earth, few have an appreciation of the geologic processes that have created it. An understanding of the Earth and the processes that are acting upon it is essential to modern man, as he struggles to make wise land use decisions under the pressures of increasing population and advancing technology. This publication, *Roadside Geology of Virginia*, by Dr. Keith Frye, serves the layman, the student, and the professional geologist by guiding him to the more important geological sites in the state, where the Earth and the processes that formed it may be studied.

Robert C. Milici
Virginia State Geologist

Table of Contents

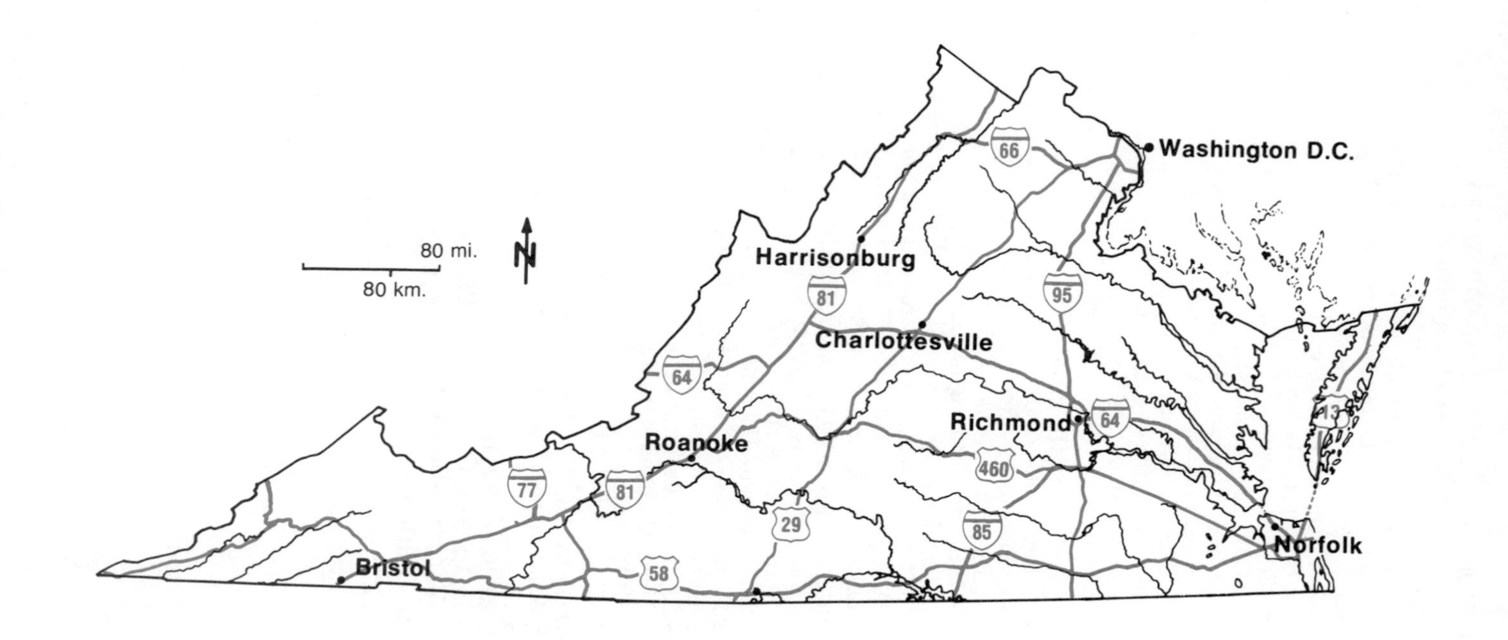

Virginia's major rivers, estuaries, and bays in black with the main highways in color.

PREFACE

Why is that hill? With *Roadside Geology of Virginia*, I hope to give you an answer to this question. And even where no hill rises to view, I try to tell you a story about the history of bedrock beneath pavement you are driving along and scenery you are looking at while traveling. Only some rocks in Virginia stand out in stark relief above roadways, but all have a story to tell you who are curious about landscapes around you.

For hidden beneath a cover of vegetation and soil is bedrock and each piece of that bedrock has a geologic history. But most soil is residual in that it represents a decay product of bedrock residing at depth. Hence, even dirt in a fresh excavation can tell you something about the history of ground around it. And that history, as you travel highways and interstates in Virginia, is what this book is all about.

Geologic features to be seen in Virginia are as varied as any in the country. Indeed, in 1985 the highway east of Natural Bridge was identified as the most geologically interesting 25 kilometers of roadway in the southeastern United States and one of the four most interesting in the country. In addition to Natural Bridge, you can see caverns still developing their unique architecture, geologic structures developed at the end of the Paleozoic era, fossils of Paleozoic life, preserved beaches from late Precambrian shores, diabase dikes, and metamorphosed Precambrian soils that developed on Grenville-age granite gneisses, all in a single short stretch of highway.

The broad outline of this book is geology along roads of the state from east to west and from north to south. Even if you are traveling in the opposite direction, some care has been exerted so that you can read each roadlog from back to front, paragraph by paragraph. Of course, you can read maps that accompany each section in any direction you choose.

Interspersed with these roadlogs, you will find short notes about geological fundamentals of each region of Virginia. These notes stand alone and may be read (or not read) in any order you find interesting.

Both author and publisher of this book about geology exposed or hidden along roadways of Virginia are firmly convinced that readers need no special background or knowledge in order to appreciate its geology and geologic history. Curiosity and interest will suffice.

Like any other science, or any specialized activity of any kind, for that matter, geology has its own specialized vocabulary of unfamiliar words and familiar words used in an unfamiliar sense. Much arcane jargon has been omitted from this book and the argot retained is explained in a glossary.

Geologists talk among themselves about 50-50 geology—observations made at 50 miles per hour of rock outcrop 50 yards away—and 50-50 geology is the only kind you can do on interstates where stopping for anything but an emergency is forbidden by law. But almost every mile of interstate highway has a parallel federal or Virginia state highway along which you can stop and inspect rocks exposed along the right of way. Indeed, many observations recorded in this book were made along highways parallel to interstates for just that reason.

So, get off your interstate occasionally and try the access road for a while. Travel may be a bit slower, but there is more to see and you can stop and inspect roadcuts at your leisure.

Maps in this book show only major roadways and towns in order to leave space for as much geology as possible. For this reason you should have a state highway map with you when you use this book to find out about geology along any particular stretch of road that interests you. And nothing stops you from adding your own roads and towns to these maps.

Sources of information used in preparing this book are too

numerous to list. They include the writings of just about everyone who has studied and published on geology in Virginia. Of special note, however, are geologists at the Virginia Division of Mineral Resources, many of whom gave unstintingly of their time and expertise during preparation of this book. In addition, Jerre Johnson and Pam Peebles of the College of William and Mary provided the geologic information for maps covering the Coastal Plain geologic province.

Any geologic map is but a progress report. Rocks do not change, but ways that geologists interpret these rocks do change as areas are mapped in ever greater detail, as new analytical techniques become available, and as new ideas enter the science. Maps in this book are as up to date as published (and some as yet unpublished) geologic data permit. Revisions of interpretation will be published, however, and you may add these revisions as they become available.

For those of you wishing additional information about any topic covered in this book, I included a list of additional readings and other sources of information. For detailed information about any locality in the state, the Virginia Division of Mineral Resources is probably the best single source.

But enough of words. Get out and let the rocks of Virginia tell their story.

Keith Frye
November 4,1985
Tyro, Virginia

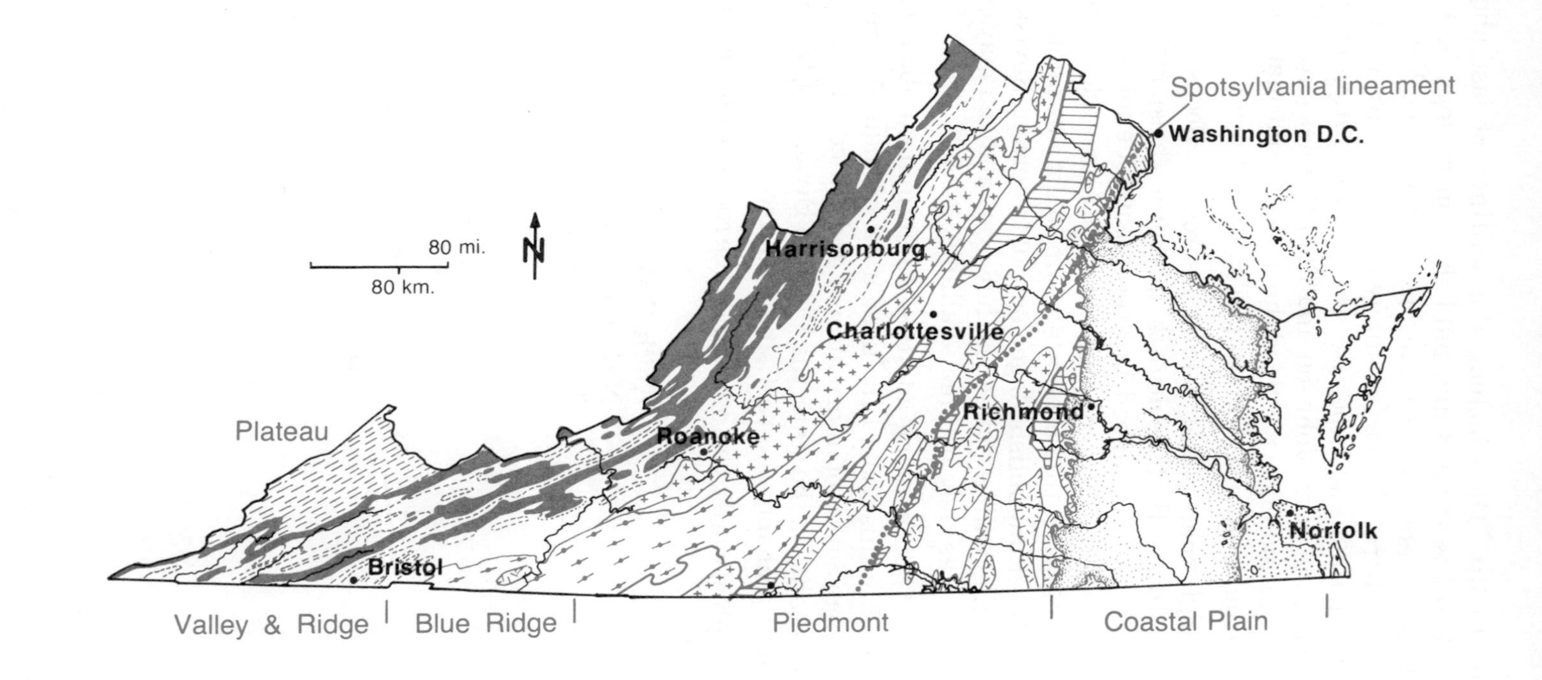

Virginia's major rivers, estuaries, and bays in black with a summary of the geology in color.

I
THE BIG PICTURE

Virginia geology cuts a small segment from the great Appalachian chain of mountains that extends from the forested ridges of Alabama to the Lapland tundras of Scandinavia. The rocks range in age from billion-year-old Grenville granites and their even older host gneisses to the modern shifting sands of Virginia Beach and the Atlantic barrier islands of the Eastern Shore. Their substance ranges from boot-sucking ooze of Coastal Plain marshes to flinty crags of Tuscarora sandstone that rise above the timber as the Devil's Backbone north of Monterey.

THE STAGE

Geologists divide Virginia into five geological provinces. 1) In southwestern counties, high cliffs of massive, horizontal red sandstone beds tower above the coal seams of the Appalachian Plateaus. 2) To the east are the sharp ridges and long, broad valleys of the Valley and Ridge province. 3) The Blue Ridge, with its Skyline Drive and Blue Ridge Parkway, runs across the Commonwealth from the Potomac River in Loudon County to the Tennessee-North Carolina border in Grayson County. 4) East of the Blue Ridge lie the deeply weathered, rolling lands of the Piedmont province. 5) The Fall Line marks the boundary between the ancient crystalline bedrock of the Piedmont and the overlapping young, in some places not yet consolidated, formations of the Coastal Plain.

Each geological province has its own story, but each provincial history links with the history of the others. Ancient westward flowing rivers carried sands, silts, and muds from the Blue Ridge, Piedmont, and beyond to bury swampy forests that became coal seams, so vital to the economy of the Plateaus. Modern rivers flowing eastward across the same lands transport the sands, silts, and muds that are becoming formations on the Coastal Plain today.

The eternal hills are the stage upon which all living things act out their lives. Except for sporadic earthquakes and volcanic eruptions, they do not appear to change during your lifetime, or your grandfather's. Indeed, rock outcrops in scenes painted centuries ago are quite recognizable today. Except for roadcuts, the Blue Ridge has the same skyline today as that seen by the first settlers in the Shenandoah Valley. Willis Mountain still stands as a sentinel above the Piedmont.

Each rock became the way it is today through events that shaped and molded it. It has a history. And we can figure out that history if we are clever enough. The history of the Earth is read in its rock formations.

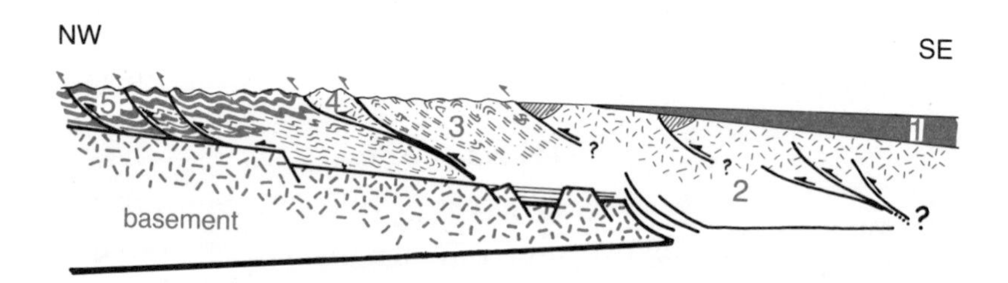

The greatly simplified and stylized cross section of Virginia geology from the Atlantic Ocean through the Valley and Ridge shows from right to left 1) Coastal Plain sediments overlying crystalline basement and Triassic basin, 2) Petersburg granite terminated by Triassic basin, 3) Piedmont shown by open fold pattern, 4) Blue Ridge crystalline rock, and 5) Valley and Ridge shown by solid fold pattern. Note that sedimentary strata extend eastward under the Blue Ridge and western Piedmont. The Triassic basin boundary faults began their history as thrust faults (arrows), but reversed their sense of movement during the Triassic.

Weathering etches out individual laminations in the Conococheague limestone of the early Paleozoic carbonate bank on US 250 near Churchville.

Geology is the science of reading history from a rock. Each geologist adds a bit to the vast body of knowledge that makes up geology today. With this book I want to tell you some of the history in the rocks you can see from the main highways in Virginia.

THE PLAYERS

Geologists read historical information from rocks, the record of events that each rock may yield upon close inspection. A limestone, for instance, tells of an ancient shallow tropical sea; a basalt, an extinct volcano; and a granite, a mountain range long since eroded away.

Igneous rocks were once molten. Melting erases a rock's history, but not completely to the careful observer. For instance, we know from laboratory experiments that basalts melted at much greater depths in the Earth than did granites. Yet basalts erupted before they froze, whereas granites crystallized deep within the crust. Igneous rocks that erupted are classified as extrusive or volcanic and those that did not as intrusive or plutonic.

If igneous rocks tell something about the interior of the Earth in times past, sedimentary rocks tell what the surface was like. They began as sediments in seas now gone or on pre-existing lands. But even here you can see back beyond the beginning if you look closely. Titanium minerals in a modern beach sand at Virginia Beach indicate, for instance, that some of the grains, at least, came from the titanium-rich central Blue Ridge.

Sedimentary rocks are classified first as clastic (top) or chemical (bottom). Clastic rocks are further subdivided on the basis of particle size and particle identity. The symbols in color are commonly used on geologic maps and cross sections.

			quartz	quartz & rock fragment	quartz & feldspar
		symbols			
clastic rocks	coarse		quartz conglomerate	graywacke conglomerate	arkosic conglomerate
clastic rocks	medium		sandstone	graywacke	arkose
clastic rocks	fine		siltstone or shale		
chemical rocks	sandy		sandy limestone and dolomite		
chemical rocks	pure		limestone and dolomite		

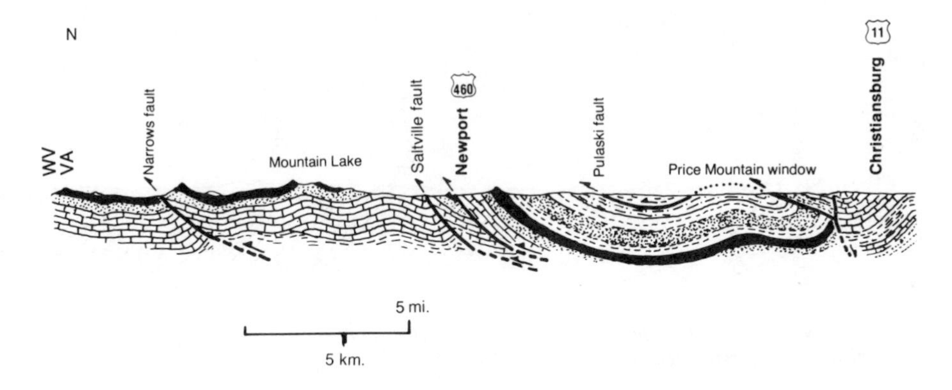

Cross section of southwestern Virginia from the West Virginia state line north of Radford to the North Carolina state line west of Danville.

Most sedimentary rocks are composed predominantly of silicate minerals, such as quartz or clay, or of carbonate minerals, such as calcite or dolomite. Predominantly silicate sedimentary rocks are further divided according to grain size—sandstone composed of visible sand grains and abrasive enough to sharpen your grandfather's axe, shale of invisibly small clay grains. Carbonate minerals are hard enough to distinguish that we will not try.

Any rock can be buried deeply enough in the Earth to be recrystallized almost beyond recognition but without actually melting. The result of this transformation is a metamorphic rock. The slates with no visible grains, schists in which micas predominate, and gneisses in which feldspars predominate of the Blue Ridge and Piedmont geological provinces all have been metamorphosed, some as many as four different times.

At one extreme, you may be challenged to recognize the pre-metamorphic character of the Candler schist in the cut along the road leading up to the Smith Mountain dam. But the rock exposed at the Greenstone Overlook at milepost 8-9 along the Blue Ridge Parkway is easily recognizeable as a metamorphosed basalt flow.

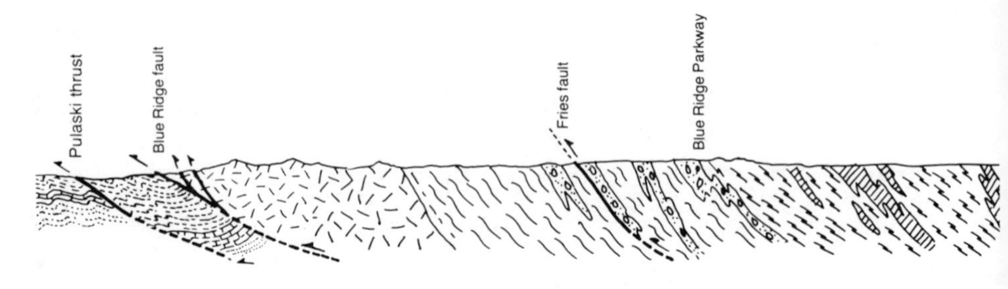

The Valley and Ridge geologic province extends from just west of the Narrows fault to the Blue Ridge fault. The Blue Ridge geologic province extends from

THE FORCES

Just look at the tortured rocks of Virginia, the sedimentary layers standing on end, the faults that moved large slabs of rock long distances. Some force has wrought havoc on the rocks of Virginia from Cumberland Gap to the crystalline basement rocks under the Virginia Beach and the Eastern Shore sands.

The sedimentary rocks of the Valley and Ridge have been folded or crumpled and faulted or broken on a grand scale. Some force shoved them tens of miles to the northwest. In addition to being folded and faulted, the rocks of the Piedmont have been metamorphosed at pressures and temperatures far beyond any that could have been created by simple burial under later sedimentary strata.

Most geologists, although by no means all, regard the theory of plate tectonics as central to explaining the condition and distribution of rocks today, from Virginia to the farthest ends of the Earth. This theory draws on data from many branches of geology.

The basic theory of plate tectonics treats the outer 50-100 miles of the Earth as rigid, rocky plates ranging in size from the Anatolian on which much of Turkey rides to most of the Pacific Ocean basin. The North American plate extends from the North Pole to Central America and from the San Andreas fault in California to the mid-Atlantic ridge, including the western half of Iceland and all of Greenland. Three kinds of plate

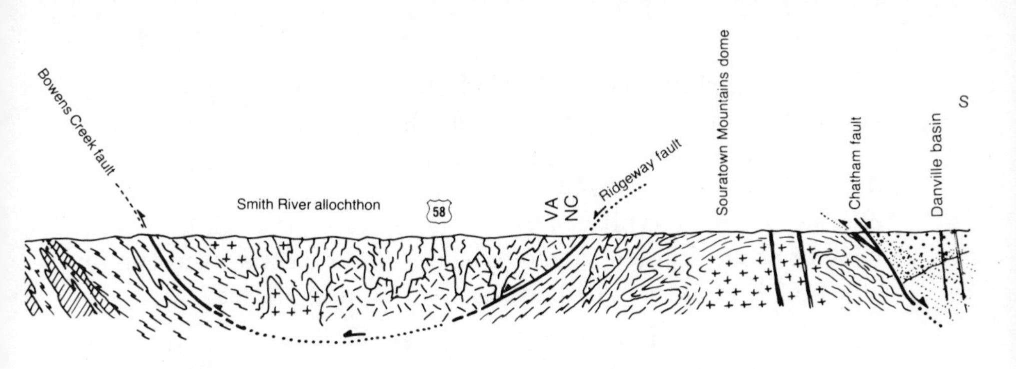

the Blue Ridge fault to the Blue Ridge Parkway on the crest of the Blue Ridge scarp. The Piedmont geologic province extends from the Blue Ridge scarp to the Danville basin.

boundaries exist—between separating plates, between colliding plates, and between plates sliding horizontally past one another.

The mid-Atlantic ridge marks a separating boundary where basaltic magma rises to fill any gaps before they can occur. Many volcanoes dot the mid-Atlantic ridge. The Andean Coast of South America is a region of colliding plates where continental rocks of the South American plate override the oceanic crust of the Pacific. California's San Andreas fault is the sliding boundary between the Pacific and North American plates.

Each continent is an assembly of fragments of small plates or microplates that became attached during some prior collision of the Andean or San Andrean type. Virginia contains at least four of these former microplates or terranes.

FIRST, GO BACK IN TIME

To appreciate the vastness of time and space involved in the assembly of rock that now supports the Commonwealth of Virginia, imagine a time machine somewhere along the Blue Ridge in the central part of the state. Imagine further that it transports you back in time to that period known to geologists as the Eocambrian, some 600 million years ago, perhaps an eighth of the age of the Earth.

You are standing on a granite headland looking east over a blue gulf toward that submerged real estate much later to become Charlottesville. Beyond the horizon is a continental island like modern New Zealand. Or a peninsula anchored to the mainland of North America in the neighborhood of Fredericksburg or Alexandria, much like Baja California is anchored in the vicinity of San Diego. Beyond that is a now-extinct ocean known to geologists as the Iapetus. And the far shore of that later will become Europe and Africa.

No tree or shrub or tall grass blocks your view, since they, too, are part of the Eocambrian's geologic future. The valley below is choked with sand, gravel, and boulders shed off this granite headland by the processes of chemical and physical weathering. The sun overhead is hot because the coastline lies from east to west somewhere near the Eocambrian equator.

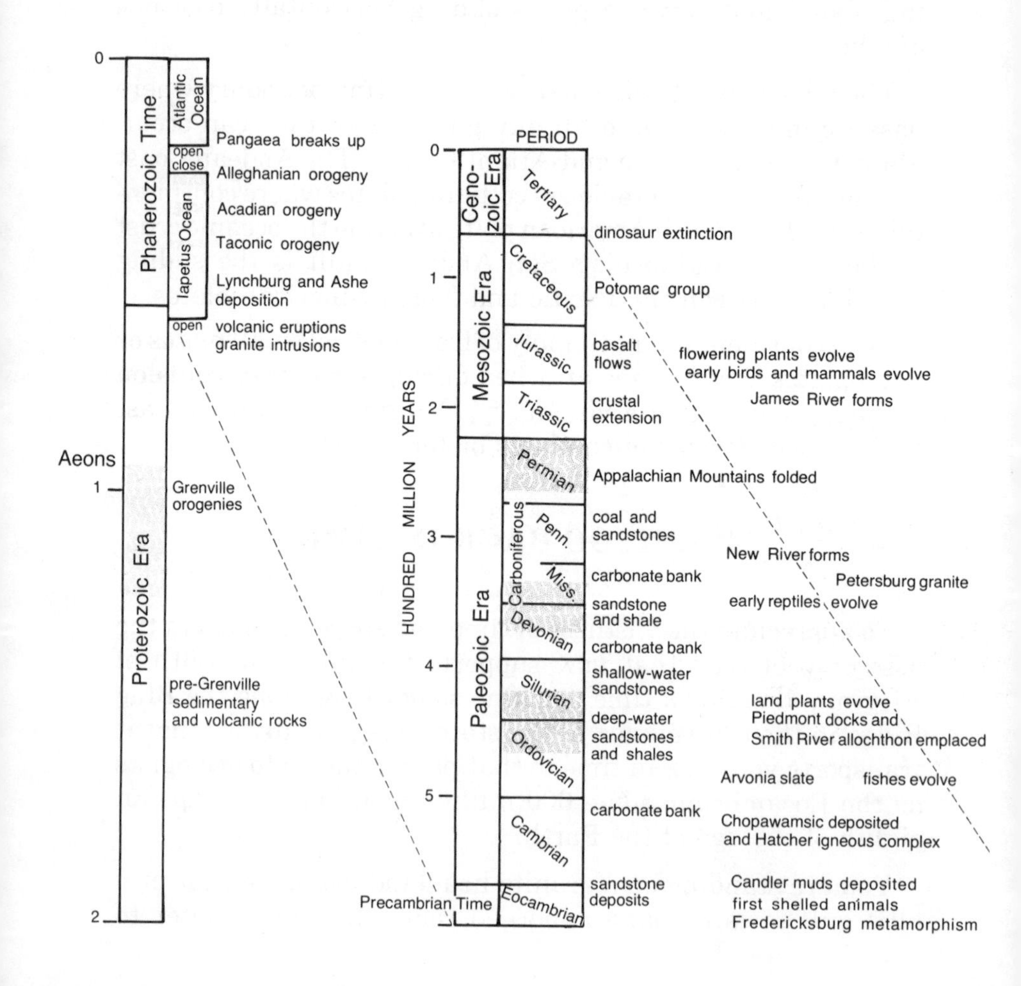

Behind, toward what will become the Shenandoah Valley, is granite and gneiss; the limestones, sandstones, and shales are yet to come. The bedrock is already old; it crystallized at depth within the crust of the Earth a half billion years before the time to which you have been transported.

Between the granite headlands are boulder-strewn beaches racked by infrequent storms much like modern hurricanes, but possibly of far greater intensity. How do geologists know this? Off shore are sand and boulder deposits to be preserved in the geologic record and named by twentieth-century geologists as the Lynchburg group of rock formations.

Reconstruction of our Earth's geologic past is an ever-unfolding, never-ending detective story. No one was there in the Eocambrian, but geologists look at the rock deposited at that remote time, assume that the laws of nature observed today applied then, and recreate the history that must have been.

Four different time scales represent the geologic history of Virginia. Times of mountain building and metamorphism (orogeny) are in color. The beginning of the Proterozoic Era at 2500 million years and the earlier Archean Era are not shown because no record of those times remains in the rocks of Virginia.

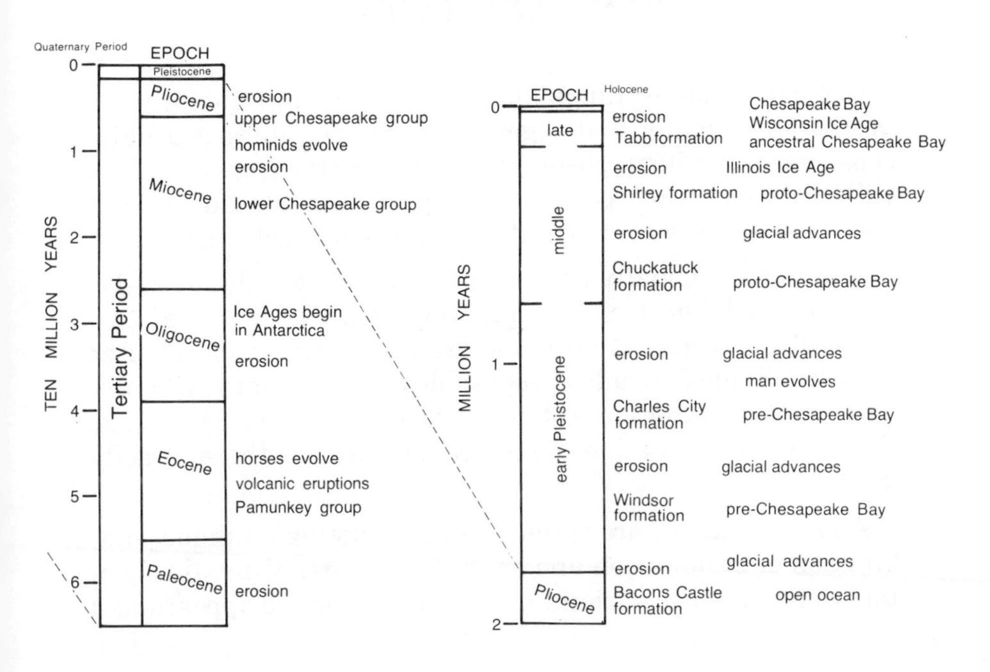

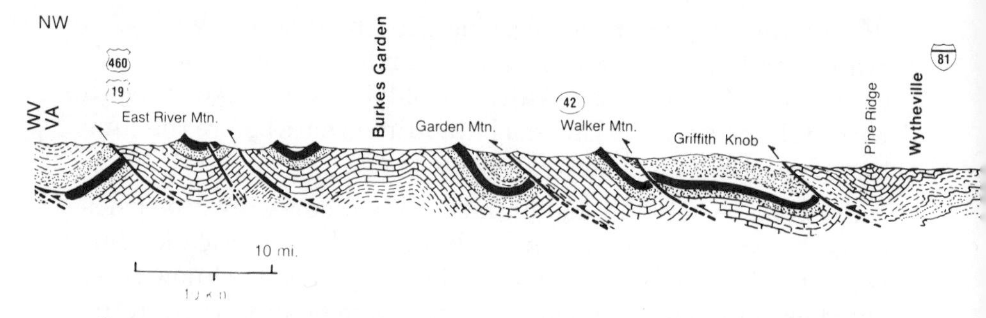

Cross section of southwestern Virginia from near Bluefield to Independence. Valley and Ridge geologic province extends from the northwest to Mount

In a band running north-northeast and south-southwest from Lynchburg are rocks that could only have been laid down on a slope much like that which connects a continental shelf with oceanic depths today. To the east are rocks normally found only in ocean basins today. Along Swift Run on the eastern flank of our modern Blue Ridge (US 33) are sandstones resembling sands today found in steep-walled valleys and bays. From these and other clues geologists can tell how our geologic past evolved.

NAME AND KNOW IT

Geologists name things and these names intimidate some people, baffle many, and anger yet others. But there is usually some reason, be it simple or mysterious, for these names. They will become a little less intimidating, baffling, or maddening if you know the reason and rules behind name selection.

A mineral is usually named by the person who discovered it and first published a determination of all its physical and chemical properties. A mineralogist may name a mineral after a person, living or dead, a geographic location, a mine where it was found, or one or more of its physical or chemical properties. The only names a mineralogist cannot use are those already used.

Approved names are species names denoting a unique combination of composition and crystal structure. Mineral collectors commonly use varietal names for distinctive appearances

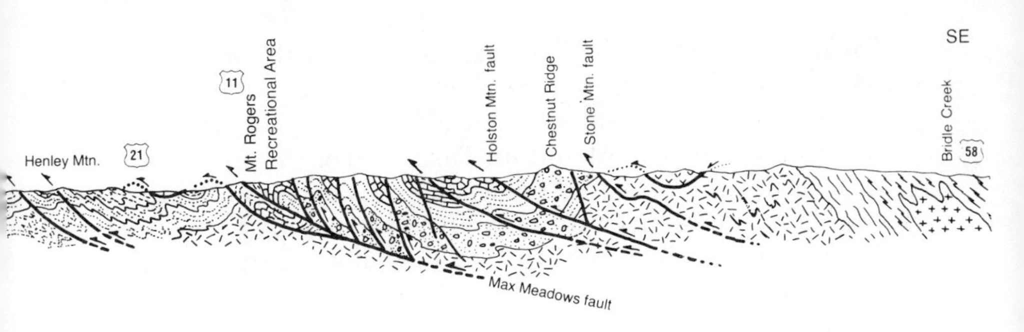

Rogers area. Blue Ridge geologic province extends from Mount Rogers area to the southeast end.

within a species. Thus, amethyst and emerald are varietal names of the species quartz and beryl, respectively. When individual mineral species are difficult to recognize without specialized equipment, geologists use group names. Garnet is a group of minerals, all having similar structures, but each being chemically different. Garnets include pyrope and almandine, to name two.

Every rock has a name. In fact, like people, it has two names. One is generic or family and the other is specific, like your first name, and both can cause problems. The generic name is sandstone or granite or some other name that tells you what kind of a rock it is. The specific name tells you which of the many sandstones or granites we're talking about.

Even after you decide what sandstone is under discussion, tracing the particular outcrop across country may reveal the character of the rock changing along strike, a facies change. Farther along, that sandstone may grade imperceptibly into a shale, then into a limestone. Rocks are like that. At any one time, different rock types are being deposited in different environments.

The specific name for a rock is a geographic name, usually a locality where the features of the rock are best displayed. The sandstone that holds up Tuscarora Mountain in Pennsylvania is the Tuscarora sandstone. And the sandstone that holds up Clinch Mountain in Virginia and Tennessee is the Clinch sandstone.

Subsequently, geologic investigation reveals that both the

Tuscarora and the Clinch sandstones occupy the same place in the geologic column. In other words, they are the same rock. After long usage in many publications, you just can't call the sandstone at Clinch Mountain Tuscarora or the one at Tuscarora Mountain Clinch. And when the sandstone holding up Massanutten Mountain turns out to be roughly equivalent to both, which do geologists call it? The Massanutten sandstone, of course.

The rock names for the geologic column in Virginia have recently been simplified to six different, but overlapping, columns for six different regions in the state. For each of the regional maps there is a single geologic column with names best fitting the formations in that region.

Rather than bog down in all the rock formation names that geologists use in their work, whole groups of similar formations may be lumped together with a time name and a generic name. Thus, these sandstones along with other related formations can come under the heading of the Silurian sandstone wedge, which happens to include a couple of other geologically related formations as well. Still cumbersome, but it does simplify the rock names and is jargon used by geologists themselves.

The names of some rocks tell you at once what they are; sandstone is a rock made of sand grains. The origins of other rock names are known only to the most knowledgeable. Charnockite, for example, a rock like granite but having pyroxene instead of mica as its dark-colored mineral, is named for Job Charnock, no geologist, but the British founder of Calcutta. His tombstone carved from this rock was later recognized as being

Ledges of Tuscarora sandstone rise above the New River at the Narrows. Rapids in the River are boulders and outcrops of this sandstone.
Photo by Thomas M. Gathright, II; Courtesy of the Virginia Division of Mineral Resources.

Iron furnaces such as this once dotted the landscape making Virginia a leading iron producer in the nineteenth century.

different from common granites. Charnockites make up a sizeable percentage of the crystalline basement of the Blue Ridge geologic province.

The use of names to designate periods of time is not unique to geology. Everyone knows what you mean when you refer to the Depression years or Elizabethan times. Geologic time periods were named long before geologists could attach years to geologic events.

Spelling variants plague geology much like they do other endeavors. When a geologist writes Allegheny, he refers to a river, a plateau, or a mountain range. For the orogeny that created that mountain range, he writes Alleghany. The Chilhowee sandstones are not named for the Virginia town of Chilhowie, but for Chilhowee Mountain in Tennessee.

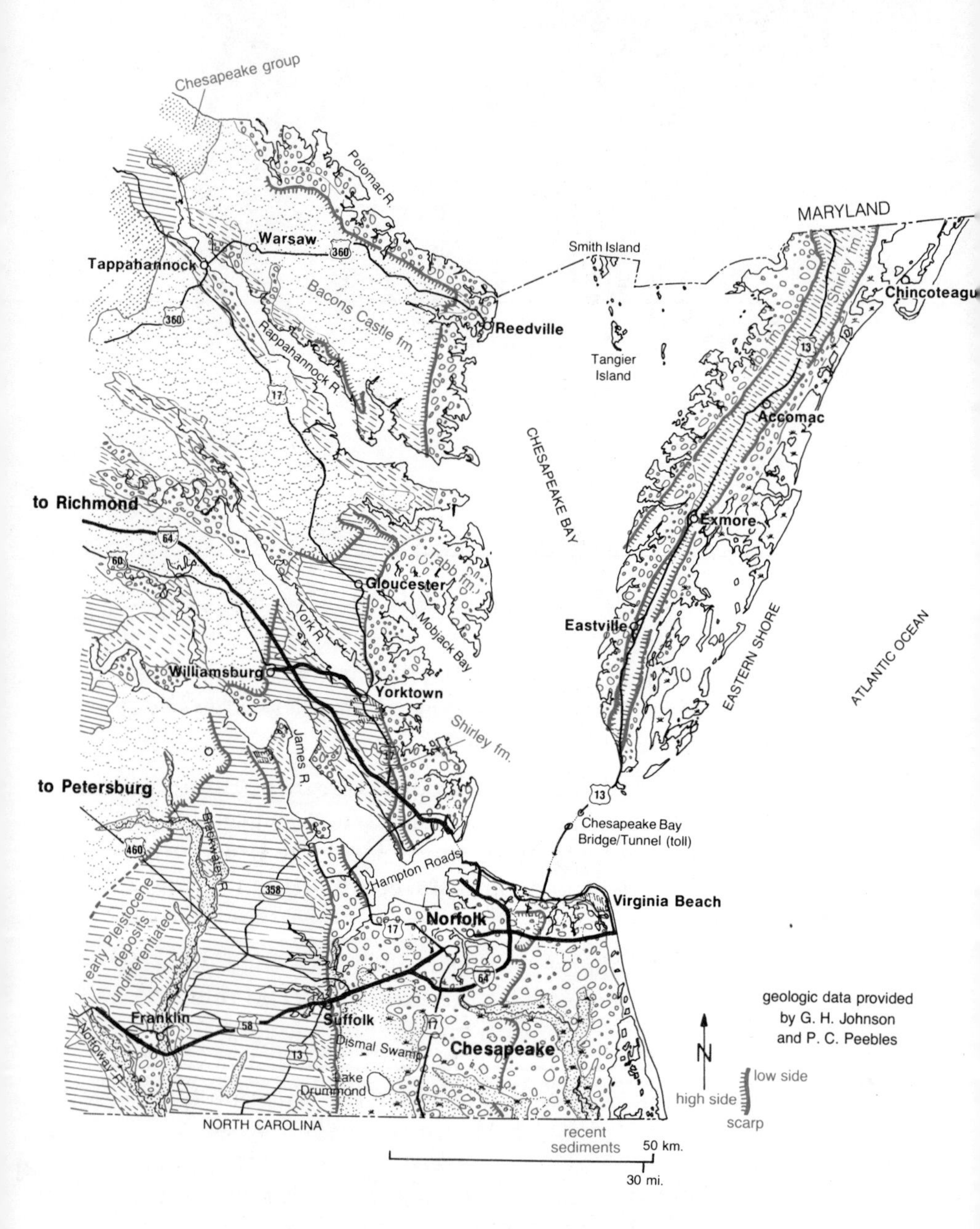

ATLANTIC AND CHESAPEAKE BEACHES

II
ATLANTIC AND CHESAPEAKE BEACHES

COASTAL PLAIN GEOLOGY AND HISTORY

The Coastal Plain surface is cut by erosion into a layer cake of sedimentary strata that rest on crystalline rock formations. These formations are similar in all ways to those exposed in the Piedmont geologic province west of the Fall Line, the western boundary of the Coastal Plain geologic province. The eastern boundary is the Atlantic Ocean, but formations of the province continue beneath the continental shelf. The shore line moves with changes in sea level. Swelling and shrinking of the mid-ocean ridges, glacial advance and melting, and the ups and downs of local tectonic activity can change the volume of the oceans, thus shifting the shore line.

Rifting that eventually opened the Atlantic Ocean began in late Triassic time, about 220 million years ago. Rifting "necked down" the continental margin much the way a stretched rod of metal necks down to a thinner diameter before it breaks. This thinning of the crust lowered its upper surface below sea level. That permitted local flooding across the thinned crust and deposition of marine sediments. In the geologic history of the Coastal Plain province, rise of sea level with flooding and deposition alternated with lowering of sea level and erosion of older sedimentary deposits.

A rising sea erodes a sea cliff and winnows out fine grained sediment. As the sea cliff retreats landward, a layer of pebbles, cobbles, and boulders remains behind. That deposit becomes a conglomerate at the base of the formation laid down during that high stand of sea level. For convenience, each high period of high sea level carries the name of the formation it laid down. Each is a period of Atlantic Ocean history. The Yorktown Sea, for instance, was the high stand of the Atlantic Ocean that laid down the Yorktown formation.

A drop in sea level exposes the former sea floor to erosion. Streams erode their channels into the recently deposited sediments and cut through these youngest sediments into older ones, even to crystalline basement rocks in their upper reaches. The modern uplands of the Coastal Plain are typical of what happens when sea level drops and the land emerges. The tidal waters of Chesapeake Bay and other estuaries, sea cliffs, bar-

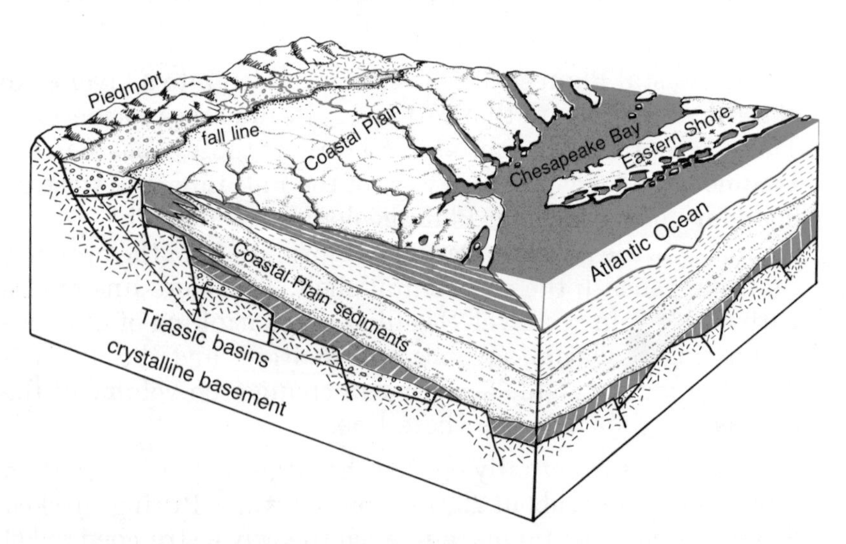

A cut-away block of eastern Virginia shows the relationship of the present surface to the crystalline basement underneath the Coastal Plain sediments. The locations of basement normal and reverse faults is schematic. Generalized Coastal Plain sediments dip toward the Atlantic Ocean.

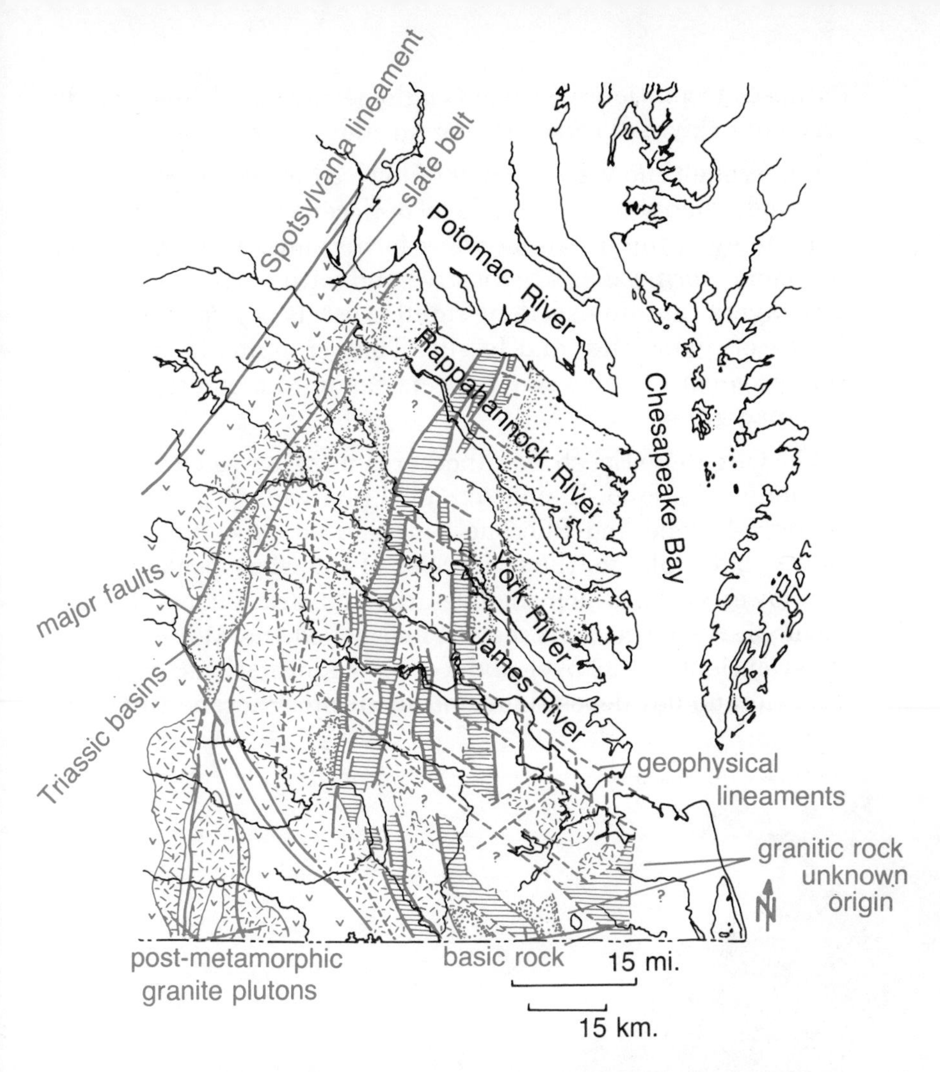

Remove the Cretaceous and Cenozoic sediments from the Coastal Plain geologic province and this is how the geology of the region might look. Some of this subsurface geology is based on samples from wells that penetrate to the pre-Cretaceous basement rocks and some on interpretation of geophysical data, such as seismic velocities and detailed measurement of variations in gravity and geomagnetism.

rier bars and islands, and off-shore bars are the result of the latest rise in sea level and submergence of the land.

The oldest known sedimentary formations of the Coastal Plain fill Triassic basins with Triassic and Jurassic sediments. Sedimentary rock deposited during Cretaceous time, the Potomac group, cover them. They include the Patuxent and

Patapsco formations—gravel conglomerate, sandstone, sandy clay, and clay. All tend to crop out along rivers.

Paleocene time was a period of low sea level and erosion of the land. Then the rising sea level of Eocene time deposited the Pamunkey group of sedimentary formations. The Aquia and Nanjemoy formations are mostly sand with local clays. Some clay layers contain excellent small crystals of gypsum, green grains of glauconite, and fossils. They crop out in the river valleys generally above and downstream from outcrops of the Potomac group.

The Oligocene Epoch was another time of low sea level—and erosion of the land. Rising sea level and widespread flooding during Miocene and Pliocene time caused depositon of sediments that make up the two cycles of the Chesapeake group. The lower cycle, which includes the Calvert, Choptank, and Saint Marys formations, began and ended the Miocene. The upper cycle, the Eastover and Yorktown formations, began in the Miocene but deposition continued into Pliocene time.

Sea cliffs along the Potomac River below Stratford Hall, colonial home of the Lee family.

Ferruginous sandstone is actually a soil horizon in which iron oxides leached out of shallow layers precipitate as cement in deep layers of sand and gravel. This sandstone is substantial enough to have been used in the construction of the mill at Stratford Hall.

Both cycles include sands, clays, and shelly coquinas. They underlie upland areas of the western Coastal Plain. At Yorktown, the cross-bedded coquinas, or fossil shell hash, of the Yorktown formation form the cliff into which Cornwallis Cave is eroded. The sea cliffs at Westmoreland State Park and on either side of the restored mill at Stratford Hall are capped by Yorktown sediments, but expose the underlying Eastover and Choptank formations. The red layer cemented by hematite, the mineral form of iron oxide, is not a formation. It is a soil layer formed as rain water dissolved iron from the overlying layers and precipitated the hematite below.

In the 1.8 million years since the beginning of the Pleistocene Epoch, sea level rose six times to cover at least part of earlier deposits with new sediment. Five times the sea shrank from the continent to permit rivers to cut channels through these sediments to a lowered Atlantic Ocean. The last period of low sea level corresponded with the last Ice Age. Enough water was tied up in glaciers then to drop sea level some 400 feet, and move the Atlantic beaches about 50 miles east of their present locations.

No Chesapeake Bay existed for the first half of the Pleistocene Epoch. Sediments accumulated then on an open coast. They are the Windsor and Charles City formations separated by an interval of low sea level and erosion.

Middle Pleistocene deposition laid down the Shirley formation, the surface deposit under much of Hampton and Newport News and along parts of the lower reaches of the York and James river estuaries. Late Pleistocene sediment makes the Tabb formation, which geologists divide into three members on the basis of three sea-level stands recorded within it.

The rise in sea level since the end of the last Ice Age flooded the coast and caused widespread deposition of beach, estuary, and swamp deposits still accumulating in low areas. These will become formations if a rising sea level causes younger sediments to accumulate on top of them. If sea level falls, they will be eroded away.

Topographic relief is low over most of the Coastal Plain and most of the sediments are so unconsolidated they can be dug out with a backhoe. Very little rock is visible except in sea cliffs and recent excavations. Most geologic studies of the formations of the Coastal Plain geologic province are done in commercial borrow pits and from drill cores.

Cyprus trees hung with Spanish moss occupy the swamp behind the dunes in Seashore State Park at Cape Henry.

A BEACH IS MORE THAN A SAND PILE

A beach, like any other landform, is shaped by the materials it is made of and by the forces acting on it. The major shaping force is, of course, waves breaking along the shore line. Additional beach sculpting is done by the wind, especially between major storms.

Sand dunes at Cape Henry.

Because sands making up a beach are loose, a beach tends to adjust quickly to any shift in pattern of wind and waves. For any particular sea condition, a beach assumes an equilibrium shape. This adjustment to conditons may take weeks during fair weather or hours during a major northeaster.

The equilibrium shape of a beach is quite different for different sea conditions. The two major sea conditions along the Virginia coast are fair weather with gentle to moderate waves from the southeast and storms with large waves from the northeast.

During a northeaster, the equilibrium shape is steep on the

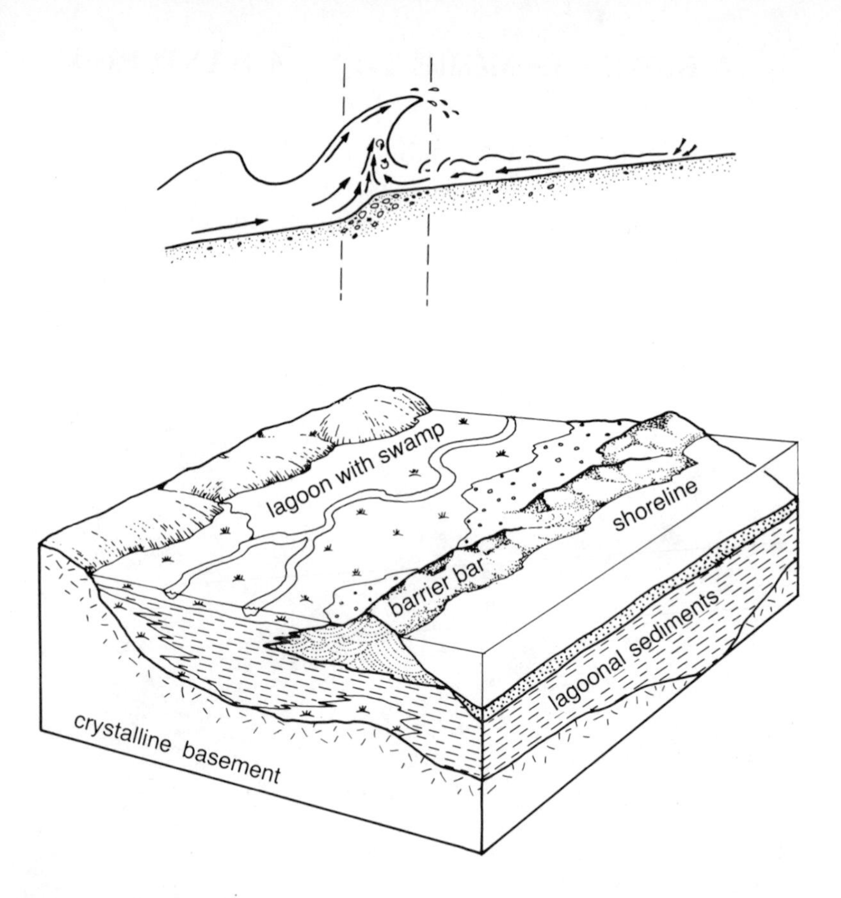

Waves coming on shore at any angle, drag on the bottom and break and fall forward, carrying sand with their flow. Waves then recede and carry the sand seaward. Virginia's Atlantic beaches typically consist of a barrier bar of cross-bedded sands (dots) piled on top of old lagoonal sediments (dashes) and swamp deposits (plant symbol).

shore side and gentle on the sea side of the water line. This profile spreads out the area of dissipation of wave energy and prevents most waves from breaking over the beach ridge and onto whatever is behind the ridge. The beach profile formed during a storm is the most efficient shape for dissipation of wave energy.

Fair weather shifts sand from shoals offshore to the part of the beach above water making the part above water wider and gentler, the shape perferred by shore dwellers and tourists alike. Most of the time the beach shape is in transition from one equilibrium shape to the other.

More wave energy impinges on Virginia Beach during the prolonged periods of fair weather, especially during the summer months, than comes from infrequent northeasters. As waves break on a beach they move sand upslope at an angle. As they wash back, the sand moves directly down the slope of the beach. So the wash of the waves moves sand down the beach. That movement is called the longshore drift.

Longshore drift on the Eastern Shore is neutral to slightly southerly. From Fort Story to Willoughby Spit on Chesapeake Bay the net drift is westward except at Lynnhaven Bay where it is eastward. From Hampton Roads north, there is no strong net longshore drift pattern along Chesapeake Bay and the Potomac River beaches.

Beaches are the boundary between land and sea. They work best when left alone and not "improved."

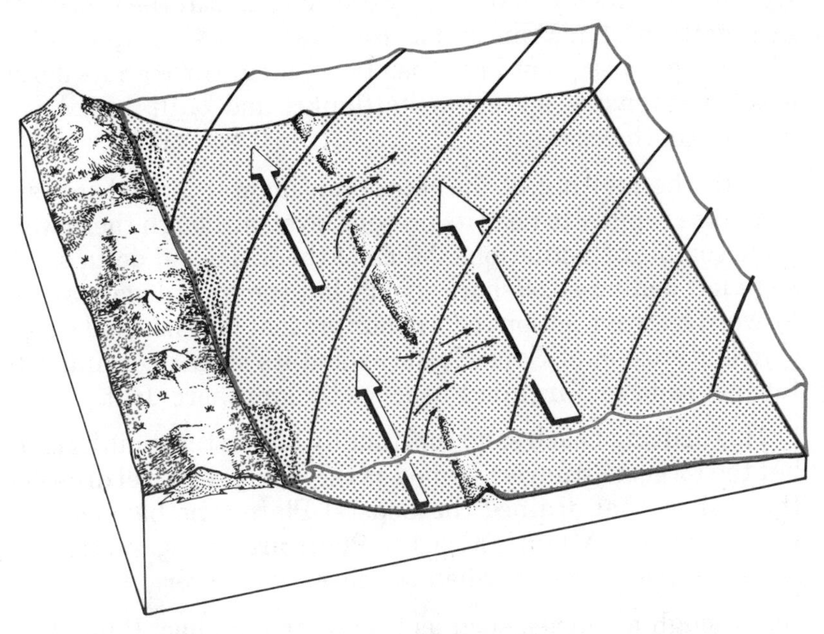

Ocean waves strike the beach obliquely most of the time. Water then splashes on the beach at an angle to the beach front tossing sand with it. As the wave recedes, sand moves directly down the beach slope. Large arrows show the net movement of this longshore drift. Breaking water tends to pile up along the shore, flow laterally a distance, and back out to sea as a rip current (small arrows). Rip currents are most dangerous when the waves are coming directly on shore.

CALL THEM RIVERS— THEY ARE STILL ESTUARIES

During the last Ice Age, the one that ended about 12,000 years ago, water from the world ocean was locked up as continental glacial ice. Sea level dropped more than 450 feet below the present level and the Virginia shore line was near the edge of the continental shelf some 60 miles east of its present location. Chesapeake Bay was then the broad valley of the Susquehanna River which continued southeastward across the present continental shelf to a shore line about 40 miles east of present Cape Charles.

At the onset of the last Ice Age, about 35,000 years ago, sea level dropped gradually and the ancestral Susquehanna River deepened its channel to match the drop. Sand and mud it scoured from its bed and banks were deposited at the shore line in a delta perhaps much like the present Mississippi delta. With each subsequent fall of sea level, it cut through its delta, leaving extensive levees on its flanks, and built a new one farther southeastward.

With the melting of the latest continental glacial ice, sea level rose rapidly to near its present level, although there were some complex fluctuations. Much of the levee and delta sands were left behind as offshore shoals, but some sand was carried forward by wave action to form our present beaches and capes. Current erosion of Cape Hatteras and its shoals continues to furnish sand to Virginia's beaches by longshore drift.

This drifting beach sand continually blocks the mouths of all but the largest estuaries formed as the rising sea level drowned the valleys that drained the Coastal Plain. The bays and estuaries of the Atlantic Coastal Plain are, thus, vestiges of ice-age stream erosion when sea level was lower.

Although estuaries such as the lower Potomac, Rappahannock, York, and James look like rivers and are named as though they were rivers, they are estuaries with important differences from true rivers. Water in a true river flows continuously downstream, but water in an estuary sloshes to and fro with every rise and fall of the ocean tides. Sewage dumped

Old Cape Henry lighthouse.

in a true river is carried downstream and away. Sewage dumped in an estuary sloshes back and forth with the changing tides.

River water is fresh from end to end, but estuarine water is fresh at one end, salty at the other, with varying degrees of brackishness in between. At a certain degree of saltiness, mud that is carried in suspension by fresh water suddenly settles to the bottom. Abnormally low or high rainfall in the watershed feeding an estuary can shift this deposition zone up or down stream.

Properly managed estuaries and marshes can be the most productive land in the world in terms of pounds of food per acre. But they are temporary and fragile features of the Virginia coastlines, accidents of the whim of sea level.

FORMATIONS AROUND HAMPTON ROADS

The geologic history of the lands around Hampton Roads is some of the most recent in the state. None of the formations that appear in excavations or in sea cliffs could have existed before the opening of the Atlantic Ocean at the end of the Triassic. Before that happened, the region occupied high ground in the central part of the supercontinent Pangaea and was shedding sediments, possibly westward.

All the formations in the Hampton Roads area were laid down within the last 5 million years in streams, swamps, or shallow seas at the continental margin. The oldest, the early to middle Pliocene Yorktown formation, crops out in sea cliffs at Yorktown and underlies the Colonial Parkway between Yorktown and Williamsburg.

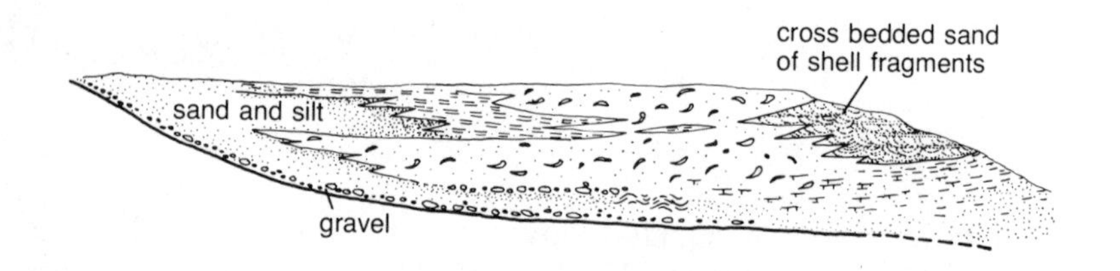

Any sedimentary rock formation will show facies variations, that is, different kinds of sediment will be deposited in different kinds of depositional environments. The Yorktown formation shown above grades upward from a basal conglomerate into sands, silty sands, and, at Yorktown, cross-bedded shelly sand. Both the sedimentary rock of the Appalachian plateaus and the Valley and Ridge geologic provinces and the metasediments and metavolcanics of the Piedmont geologic province show facies variations.

At Cornwallis Cave in the seacliff at Yorktown, you can see that this formation is made of shell fragments broken by wave pounding after the inhabitant of the shell died. This coquina is primarily the mineral calcite which may dissolve slowly to form caves as ground water percolates through it. Stories abound relating the sudden appearance of 15-foot holes in people's lawns and ditches becoming apparently bottomless right in front of a backhoe operator. Caves tend to collapse like that when calcite has been dissolved out.

Above the Yorktown formation, geologists recognize a similar sedimentary deposit, the late Pliocene Chowan River formation. It does not crop out at the surface in the Hampton Roads area, but appears in some borrow pits and in drill cores. Next up is the middle Pleistocene Shirley formation. Its lowest strata contain pebbles to boulders which geologists recognize as having come all the way from the Blue Ridge. It also contains peat from buried swamp deposits, some with tree stumps still standing. This formation is the layer beneath the soil around Suffolk and Smithfield and along the lower reaches of the James and York rivers.

Most of the surficial deposits between the Suffolk scarp and the recently active beaches belong to the late Pleistocene Tabb formation. Geologists divide it into three all but indistinguishable members; each represents a change in sea level. Currently sand moves along the modern beaches and silt and clay are raining out of the estuaries of the region.

The Cape Henry lighthouse guides ships to the entrance of Chesapeake Bay.

Wind across beach sands creates small ripples which can become buried and preserved in sandstone.

US 13
Maryland state line — North Carolina state line
123 mi./198 km.

The Delmarva Peninsula is a geographic, geologic, and economic region politically fragmented among the states of Delaware, Maryland, and Virginia. The Maryland and Virginia portions of the peninsula are also called the Eastern Shore. The two sides of the peninsula are called Bayside and Seaside. The highway and railroad mostly follow high ground, 35-45 feet above mean sea level, which is bordered on both sides by erosional scarps cut by wave action at a higher stand of sea level.

The high central part of the peninsula is underlain by barrier and beach sand deposits of middle and late Pleistocene age. On the Seaside, Assateague is a barrier island extending south from the Maryland state line. South of Assateague, small barrier islands and salt marshes, both inaccessible by automobile, front the Atlantic. More salt marshes separate the barrier islands from the mainland.

On the Bayside north of Exmore, salt marshes face Chesapeake Bay. Farther south, you can drive to beaches that face the bay. Chincoteague, Onancock, Wachapreague, Oyster, and Cape Charles are fishing ports.

During Pleistocene time, advancing glacial ice on the continents to the north captured water from the world ocean and caused sea level to drop at times as much as 400 feet below its present level. During

The harbor at Onancock shelters fishing and passenger vessels on the Eastern Shore.

interglacial periods, sea level rose as much as 100 feet above its present level building barrier beaches and eroding sea cliffs now preserved as scarps.

During ice advance and sea regression, Chesapeake Bay became the broad lower end of the ancestral Susquehanna River valley with its mouth near the edge of the continental shelf. Atlantic Coast beaches were then far east of their present locations and the present marshes were wooded valleys.

When the sea level rose as the glaciers melted, these valleys drowned to become bays and estuaries, partially filled with sediments both from the land by stream transport and from the sea by wave transport. When sea level rose to 50 feet above its present stand, the entire Eastern Shore was a long sand bar between a much larger Chesapeake Bay and the ocean. At a lower stand, waves lapped at the base of bordering escarpments, one of which you can clearly see on the west side of US 13 around Cheriton. The location of the present shoreline is but a happenstance of time and climate!

The Pleistocene plains and scarps–steps and risers–rise from Chesapeake Bay on the right to the Fall Line on the left. Each scarp represents shoreline of the bay, the oldest on the left and the youngest scarp just above the present sea level.

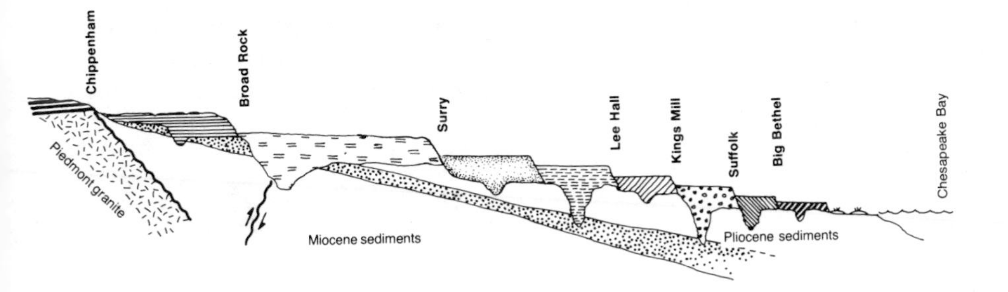

Curving sandy ridges 3-8 feet high are marked by the locations of old roads, houses, and cemeteries for most of the length of the peninsula. From the highway you can see one sand ridge 200 yards south of the junction with highway 676, just south of the turnoff to Hopeton, on the north side of Mapp, and again two miles south of Onley. Others are quite distinct as complete circles on aerial photographs. These features are called Carolina Bays.

Carolina Bays range from scores of feet to several miles in diameter. Their rims are very sandy and well drained while their interiors are swampy and hold water after rains. Several ideas about the origin of Carolina Bays have been put forth over the years since they were first recognized. Most likely they started life as dune fields drowned by a transgressive sea. Erosion and deposition by wind and water connected the dunes in a circle; then vegetation after sea regression protected them from further erosion.

Early settlers used the well-drained rims of Carolina Bays for their roads, homes, and cemeteries. Farming has destroyed the topographic expression of many of them over the years, but the sandy soil in the fields is still distinctive and visible from the air.

The Chesapeake Bay Bridge-Tunnel is a 20-mile-long bridge interrupted by tunnels under two ship channels. It was completed in 1964 at a cost of $200 million and replaced ferry service across the mouth of the bay that had run since Colonial times. To build the four artificial islands and to support and protect the two tunnels, nearly two and a half million tons of granite, sand, and gravel were barged in at a cost of $10 million.

The route crosses the Diamond Springs scarp onto the upper Pleistocene Tabb formation just south of its intersection with US 60 in Virginia Beach. Between Portsmouth and Suffolk, it crosses the north end of the Great Dismal Swamp, a large wetlands stretching south to Elizabeth City, North Carolina. Although George Washington wanted to drain the swamp, it is now valued as a nature preserve for native wildlife that commingles northern and southern species and as a staging area for migrating waterfowl.

At Suffolk, the route climbs the old shoreline of the Suffolk scarp onto the Shirley formation of middle Pleistocene age. Southwest of Suffolk it traverses the Isle of Wight plain, underlain by the Chuckatuck formation of early Pleistocene age, to the North Carolina state line.

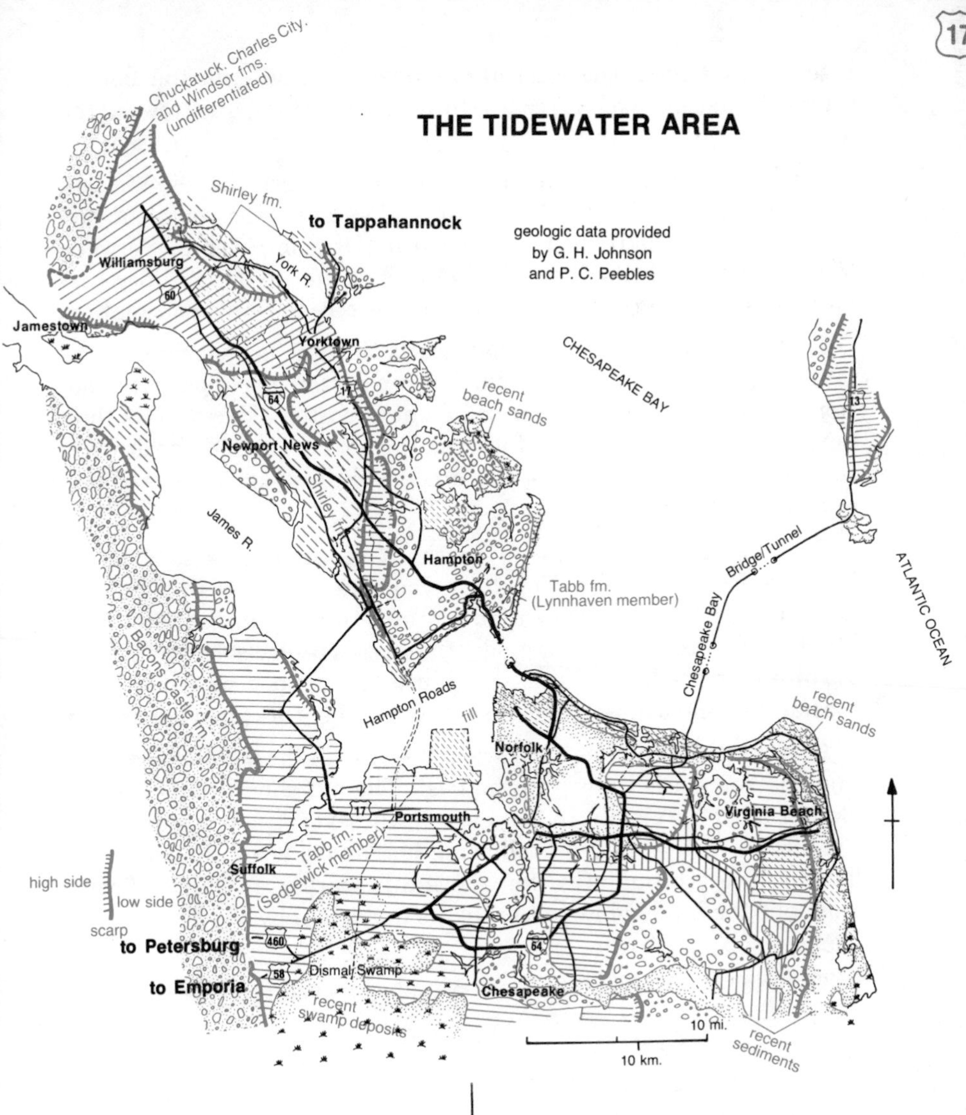

US 17
Tappahannock — Chesapeake
126 mi./203 km.

Tappahannock is on the late Pleistocene Tabb formation. It is a clayey, silty, sandy unit with zones of pebbles, fossils, and buried swamp remains. Just south of Tappahannock, near the north bank of

Piscataway Creek, the road passes down section onto the Bacons Castle formation, which is coarse toward the bottom, finer toward the top.

Between Saluda and Glenns, the route crosses the Piankatank River, which has filled with stream and swamp deposits in the last 10,000 years. Upstream it is known as Dragon Swamp. Between White Marsh and Gloucester Point on the York River, US 17 crosses along the eastern edge of the middle Pleistocene Shirley formation.

To the east at the foot of a small scarp, the land is underlain by the late Pleistocene Tabb formation. The reason for this apparent violation of the rule that younger sediments lie atop older ones is that, during the intervening time interval, the next to last Ice Age dropped sea level and this scarp is the eroded bluff of an ancestral Susquehanna River.

The sea cliffs at Yorktown on the York River are held up by Yorktown coquina. The Yorktown coquina is locally well enough consolidated to have been used initially for fill in building Fort Monroe. (It deteriorated so rapidly that it was replaced by Petersburg granite.) Above the coquina, the bluffs are capped by early Pleistocene silts and sands, which also underlie the Yorktown battlefield.

At Yorktown, the Lee Hall scarp lies just to the west of the route and the Big Bethel scarp to the east. The old shoreline of the Big Bethel scarp is visible to the east down several side roads in the vicinity of Grafton, US 17 following just above it for several miles. The highway through Grafton above this scarp is on Shirley formation, but the land below to the east is underlain by the Tabb formation.

The Pleistocene Tabb formation underlies the road on the south

The Pliocene Yorktown formation contains cross-bedded deposits of shell fragments. Cornwallis Cave at Yorktown is said to have housed supplies for the British army during the American Revolution.

side of the James River and much of the city of Portsmouth. South of Portsmouth, the route follows the Dismal Swamp Canal section of the Intracoastal Canal southward to the North Carolina state line. The swamp deposits on either side of the road are recent. Canoes may be launched at Arbuckle Landing to take the Feeder Ditch to Lake Drummond, one of two natural lakes in Virginia.

The Intercoastal Waterway provides sheltered passage for boats traveling along the Atlantic Coast.

Man is a geologic agent in beach formation. Upper left is beach improved by groins and lower left is beach bulldozed into shape with dredged sand brought down the beach by pipeline.

US 60, Interstate 64
Virginia Beach — Richmond
109 mi./175 km.

At its terminus in Virginia Beach US 60 is on recent beach deposits and it stays on them to the north end of the Hampton Roads Bridge Tunnel. East of the interchange with US 13, the Diamond Springs scarp appears on the south side of Shore Drive.

The route crosses a belt of late Pleistocene Tabb formation between Strawberry Banks and J. Clyde Morris Boulevard. From the westbound lanes of Interstate 64 about one half mile east of the interchange with US 17, you can see into the southeastern corner of the Williams borrow pit, one of a dozen or so sources of fill dirt in Tidewater. The bottom of this pit is below sea level; seepage and rain water are continually being pumped out so that it can be worked.

Excavation in this borrow pit has penetrated from the surficial Tabb formation of the upper Pleistocene into the middle Pleistocene

Cyprus stump growing on coastal swamp was buried by the middle Pleistocene Shirley formation and exhumed in place in the Williams borrow pit.

Shirley formation and still older Yorktown formation. Cypress and other tree stumps that have not moved since they lived here have been exhumed in this pit.

The Harpersville scarp is just west of the pit near the junction with US 17 and the Big Bethel scarp is just to the east. Each scarp was eroded by wave action when it marked high sea level stands of the old Atlantic Ocean. East of the scarps the route crosses the Shirley formation. West of Lightfoot and Camp Peary the underlying rock is Pliocene Bacons Castle formation.

Between Newport News and Richmond, the routes are underlain by flat-lying Tertiary sedimentary strata. The Chickahominy River, where they cross it, is distinctly flat-bottomed and swampy. The stream is underfit in the sense that the flood plain is wider than can be accounted for by ordinary stream erosion processes. Pleistocene sea level stands up to 400 feet lower than at present permitted the river channel to cut correspondingly deeper. What you see is just the upper portion of a once deeper valley now filled with swamp debris piled in during the last 10,000 years.

The James River channel is cut into the Petersburg granite at the Fall Line, but the surrounding uplands are capped by Tertiary sands and gravels.

US 460
Norfolk — Petersburg
76 mi./122 km.

The route begins on the recent sands of Chesapeake Bay in Norfolk's Ocean View section. Much of the route through Norfolk and Chesapeake is on late Pleistocene sediments and the artificial land upon which these cities are built. Drainage ditches along the highway in Suffolk mark where the route crosses the Great Dismal Swamp.

Vigorous streams choked their channels with coarse-grained bedload making the overbank saprolite (chemically rotten rock) easier to erode than their channels. Lateral migration of the stream into this saprolite then left the former rocky channels as the stream divides. Where former channels become the divides and former divides become channels, the topography is described as inverted.

The route crosses these upland surfaces and then drops rather sharply into flat-bottomed swamps of a variety of widths from a few yards to hundreds of yards. Regardless of swamp width, each side is bounded by a fairly distinct bank leading up to the upland surface. Most of these slopes have about the same angle.

The Appomattox River cut its valley through the bordering upland Tertiary sands and gravels to the underlying Petersburg granite for about 13 miles west of Petersburg. Downstream, the river cut into the Cretaceous Patuxent formation, a sandstone with interbedded clays and gravels. In Petersburg the route crosses Eocene Aquia and Miocene Calvert formations, both sandy deposits.

III
POTOMAC REACHES

GEOLOGY ALONG THE POTOMAC

The Wissahickon terrane is bounded on the east by the onlap of Coastal Plain sediments, on the west by the eastern border fault of the Culpeper basin of Triassic to Jurassic age. Arbitrarily, we truncate this terrane at the Virginia state line in the Potomac River valley in the north and the Occoquan River in the south although it clearly extends farther in both directions. It also extends an unknown distance east underneath the younger rocks of the Coastal Plain.

This terrane, like most in the Piedmont province, is buried beneath a thick layer of saprolite, weathered rock, which in turn is, in places, covered by terrace deposits. Although outcrops appear in some streams, most of the bedrock is invisible to geologists and casual observers alike. However, diligent search and careful study of the available outcrops gives us a glimpse of the complex Wissahickon terrane.

The oldest formations belong to the Annandale group of Eocambrian or Cambrian time. The rocks crop out in a window

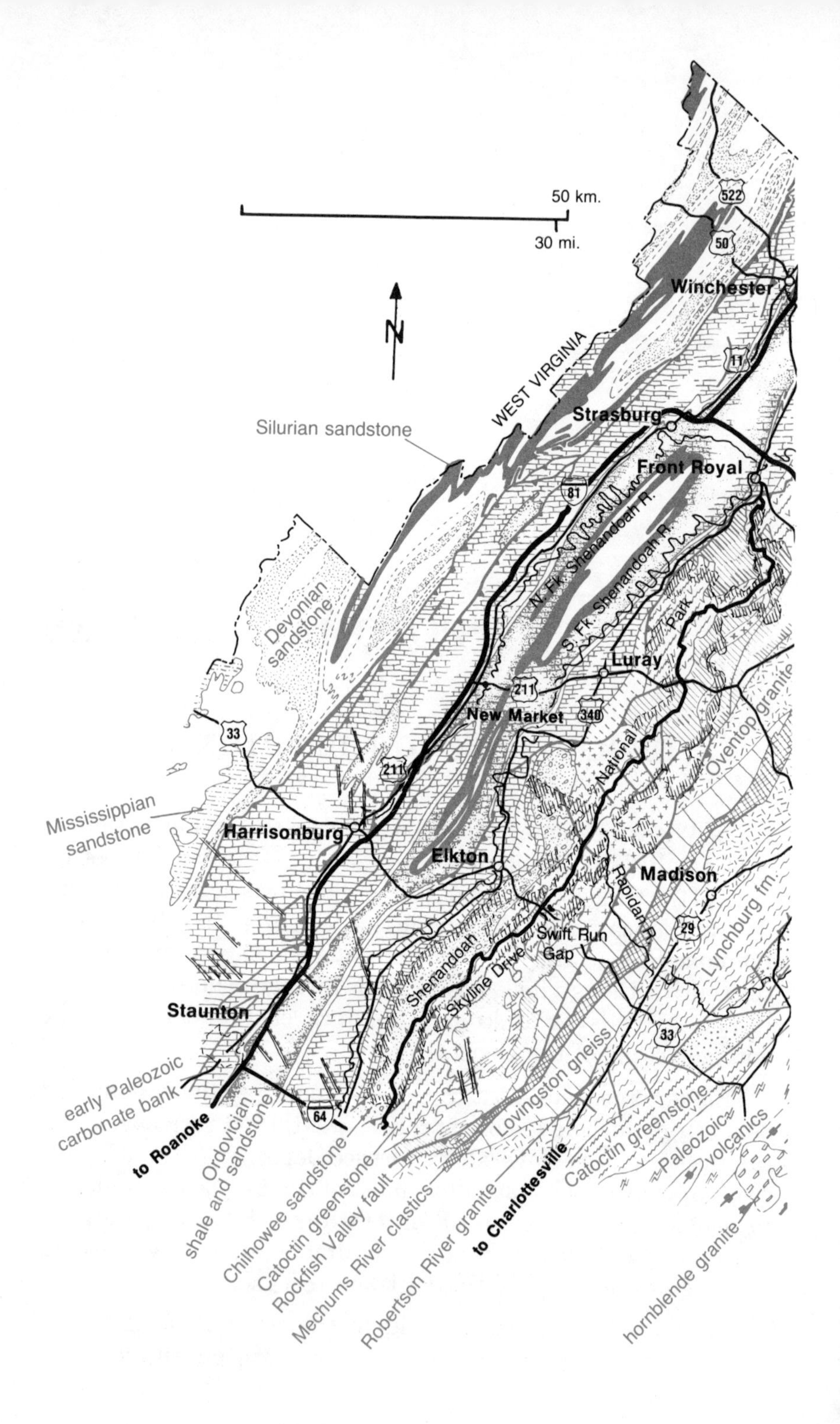

50 km.
30 mi.
N
WEST VIRGINIA
Silurian sandstone
Devonian sandstone
Mississippian sandstone
Winchester
Strasburg
Front Royal
Luray
New Market
Harrisonburg
Elkton
Madison
Staunton
522
50
11
81
211
340
33
29
64
N. Fk. Shenandoah R.
S. Fk. Shenandoah R.
Shenandoah National Park
Skyline Drive
Swift Run Gap
Rapidan R.
Oventop granite
Lynchburg fm.
early Paleozoic carbonate bank
to Roanoke
Ordovician shale and sandstone
Chilhowee sandstone
Catoctin greenstone
Rockfish Valley fault
Mechums River clastics
Robertson River granite
to Charlottesville
Lovingston gneiss
Catoctin greenstone
Paleozoic volcanics
hornblende granite

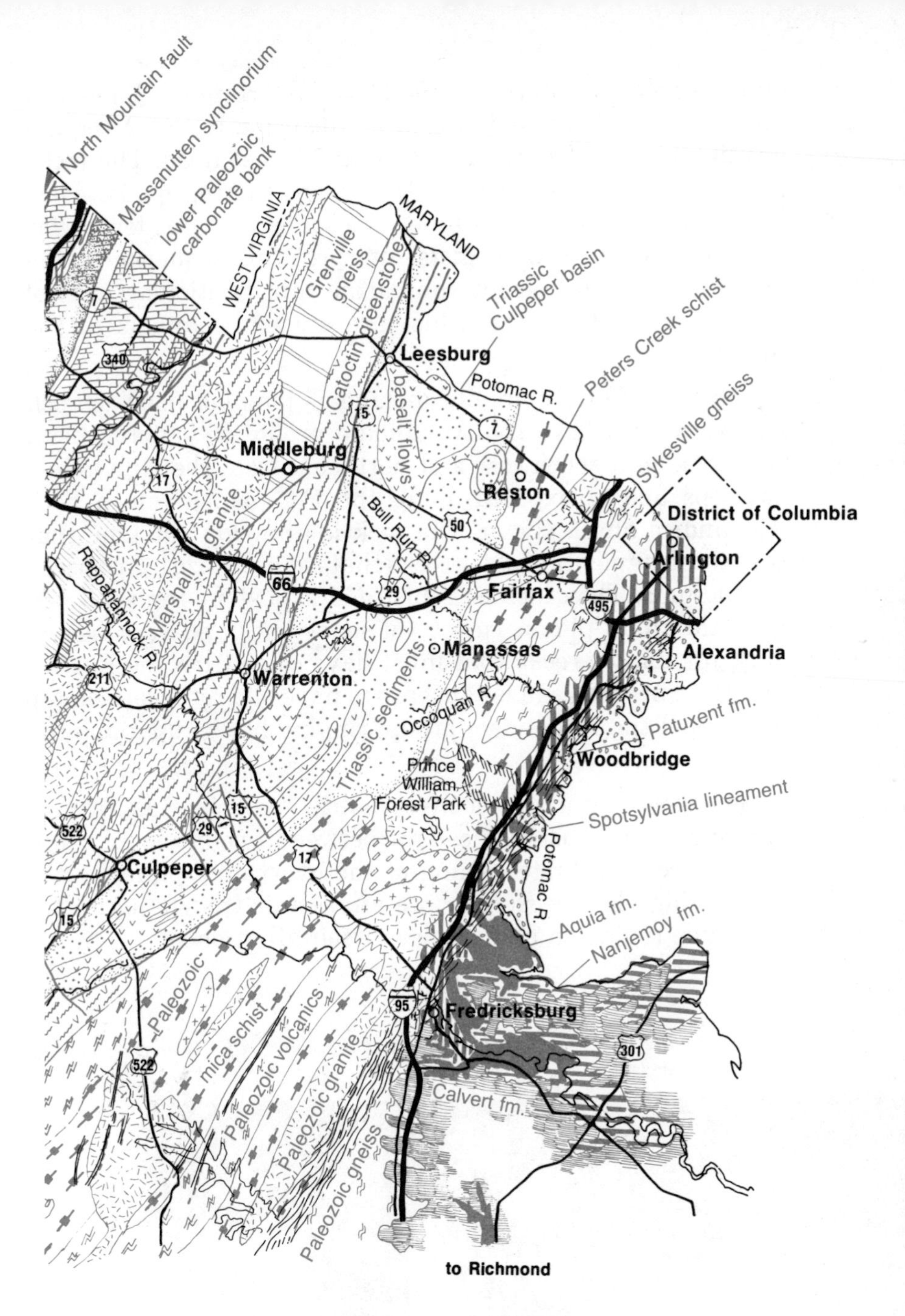

POTOMAC REACHES

on either side of Interstate 495 between Arlington Boulevard and Braddock Road and as a large inclusion in the Occoquan granite between Keene Mill Road and Hooes Road. The rocks are a lower fresh gray to yellowish weathering quartz-mica schist and an upper sandstone. The schist contains lenticular beds of sandstone. Geologists interpret this sequence as the outer edge of an ancient submarine fan. Lenticular beds of sandstone in the lower schist are ancient channels in the muddy bottom.

The Sykesville "ophiolite melange," jumbled mass of rock fragments, is above the Annandale group. Its outcrop band stretches both southwest and northeast from the Annandale outcrops and is exposed along the Potomac River in both Alexandria and McLean. The rock is a muddy siltstone that contains fragments of older rocks that range in size from specks to stadiums. They include schist and sandstone from the Annandale group plus a variety of soapstones and very heavy, dark ultramafic rocks that must have come from the Earth's mantle. This entire mass appears to have slid into place on top of rocks of the Annandale group, ripping blocks up along the way.

Geologists interpret such ophiolite melanges as sutures where oceanic crust was caught between two blocks of continental crust. Oversteepening of the overriding plate, as the underlying plate is sliding into an oceanic trench, permitted this slide to detach and emplace itself all at once.

The Sykesville formation forms cliffs near the Key Bridge over the Potomac River. Ramesh Venkatakrishnan photo.

The conglomerate of the Sykesville formation formed as a submarine slumping in an ancient ocean.

Rocks of the Potomac River allochthon, a large displaced mass, lie above the Sykesville ophiolite melange. Its outcrop band runs from the Capital Beltway north of the Interstate 66 exit westward to the Culpeper basin and from Interstate 66 between the interchanges with US 29 and US 50 north to the Potomac River. The Peters Creek schist in this displaced allochthon is predominantly a quartz mica schist with lesser amounts of dirty sandstone. This rock was metamorphosed to a migmatite, granite mixed with metamorphic rocks, and subsequently sheared to a phyllite. Its contact with the underlying Sykesville melange is a fault.

The Piney Branch allochthon was shoved over the underlying Potomac River allochthon. Its outcrop belt extends southwest from Fairfax City. Within the allochthon, the Piney Branch complex consists of very dark igneous rocks now metamorphosed to greenish serpentinite, soapstone, and dark amphibolite full of glistening crystals of shiny black amphibole.

Rocks of the Annandale group, the Sykesville ophiolite melange, and the Potomac River and Piney Branch allochthons were folded into a syncline, now overturned to the southeast, and metamorphosed before deposition of the Popes Head for-

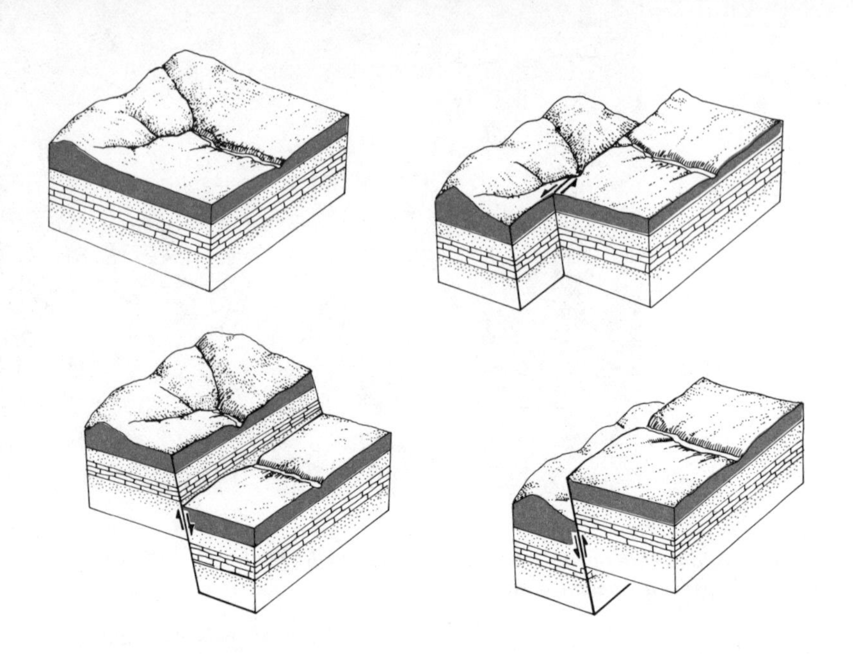

The top left block represents rock before faulting took place. Top right shows a left-lateral, strike-slip fault. Bottom left shows a normal, dip-slip fault, and bottom right a reverse, dip-slip fault. Geologists recognize these off-sets by marker beds such as the one shown in solid color.

mation, which is exposed from Vienna southwest through Clifton. The Popes Head formation is mostly metamorphosed siltstone, with lesser amounts of sandstone and metamorphosed volcanic rock. Toward the top of the formation the grain size of the original rock becomes finer and it is metamorphosed to a phyllite. The original sediments and volcanic ash falls probably accumulated in a basin between the continental margin and an island arc.

All these now metamorphosed sedimentary and volcanic rocks were later invaded by molten granitic magmas during Cambrian time, about 500 million years ago. The Occoquan granite crops out in a belt south of Little River Turnpike and west of the Fall Line. The Occoquan granite was crushed and sheared during metamorphism and, thus, converted into a streaky gneiss.

Rifting that began to open the Atlantic Ocean 220 million years ago broke the Piedmont along faults. The Triassic Culpeper basin sank and later filled with sediments and basalt flows. The Wissahickon terrane ends somewhere beneath that basin.

As the Atlantic Ocean opened, the new continental margin necked down to meet the new ocean floor, causing the sea to lap onto it. Sands deposited in and along the Atlantic in Cretaceous time became the Potomac group of formations, the Patuxent below and the Patapsco above. These formations lap onto the Wissahickon terrane along the lower Potomac.

Capping Coastal Plain sedimentary formations and the metamorphic and igneous rocks of the Piedmont are terrace deposits of uncertain age. These blanketing terraces along with the deep alteration of the crystalline rocks to saprolite have made the rocks extremely difficult to see and to interpret.

CONTINENTS AND OCEAN BASINS

Geographers distinguish continents from other terrestrial features on the basis of size; geologists make the distinction on the basis of rock type. If the underlying rock is predominantly granitic and overlain in places by thin to thick—thousands to tens of thousands of feet—sedimentary sequences, that part of the crust of the Earth is continental.

If the underlying rock is predominantly basaltic with a relatively thin—tens to hundreds of feet—sedimentary cover, this is oceanic crust. Transitional crust exists at continental margins and locally in ocean basins.

Geologic continents that are smaller than geographic continents, for example New Zealand or the Seychelle Islands, are called microcontinents. Isolated blocks of the Earth's crust that are transitional between continental and oceanic include island arcs, such as the Aleutian Islands, and oceanic plateaus, such as the Blake plateau, off the Florida coast.

Think of continental crust as analogous to pine planks and oceanic crust to oak planks, both floating in water. Transitional crust might be analogous to anything from cedar to maple. Clearly, the low-density pine planks float higher in the water than do the high-density oak planks.

Continental and oceanic plates "float" on a plastic upper zone of the Earth's mantle that geophysicists call the asthenosphere. It can flow, but more like cold molasses than like water. And low-density continents float higher above the asthenosphere than does high-density oceanic crust. Oceans are wet because oceanic crust is thinner, denser, and rides lower than continental crust, so water collects in topographically low parts of the globe. Continents are dry for the most part because they are thick enough and light enough to keep much of their surface above sea level.

This distinction between continental and oceanic crust on the basis of rock type is important in understanding the rock types and structures that make up Virginia. Oceanic crust may sink or slide under continental crust, and generally does; continental and transitional crust resist subduction because they are lighter, therefore more buoyant.

As the various plates of the lithosphere move about and collide, oceanic and continental crust behave differently along lines of collision. At a convergent, destructive plate margin—an oceanic trench—oceanic crust generally sinks into the Earth's mantle where it loses its identity as a crustal rock; it is subducted. Two continents in collision cannot destroy one another, but may pile up a double crustal thickness as under the modern Himalaya Mountains.

Microcontinents and transition zones behave in the same way as continents during plate tectonic interactions; hence geologists treat them in the same fashion as their larger brethren. The Paleozoic history of Virginia is punctuated by two episodes during which an island arc or a microcontinent collided with the eastern margin of North America and one during which major continents, Africa and Europe, changed the geologic development of the area.

Long before the causes of these punctuations in the geologic record were understood, their effects were recognized. Before the theory of plate tectonics, those effects were called orogenies—periods of mountain building. Orogeny is now understood as the effects of collision between continental or transitional crust. Oceanic crust is rarely preseved at a collisional plate boundary—it is subducted—but its effects are as profound as the Andes Mountains or the Aleutian Island arc,

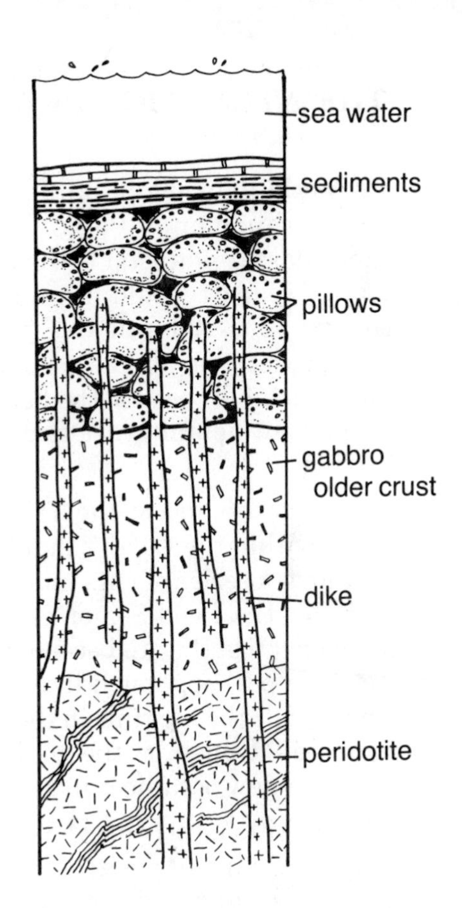

Oceanic crust commonly consists of horizontal sediments, usually containing chert, on top of pillow basalts. These keg-sized pillows result from blobs of molten magma frozen by contact with sea water. In and below the pillows, dikes of basalt intrude older crust and feed submarine vents to form the pillows. Beneath the crust is peridotite, rock composed primarily of olivine and garnet. Geologists interpret this association of rocks, when found in the Piedmont geologic province, as a suture between microplates once separated by an ocean.

where Pacific Ocean crust is subducting today.

Where bits of oceanic crust are preserved within a continent, such as in the Sykesville formation along the Potomac River, geologists can state with some certainty that oceanic crust once existed there and was mostly, but not entirely, consumed by subduction. These entrapped—obducted—oceanic rocks mark the location of a suture where two continents or microcontinents joined.

Geologists and geographers are not all that much at odds with one another: Almost all continents and continental margins are underlain by predominantly granitic rock. Almost all the ocean basins and basin margins are underlain by predominantly basaltic rock. Only where size and composition do not coincide is there any discrepancy between geographic and geologic considerations of which is which.

THE ATLANTIC OCEAN IS A TRIASSIC BASIN

A world map of 220 million years ago is dominated by one great continent—Pangaea—and one great ocean—Panthalassa—ancestor to the modern Pacific Ocean. Virginia's nearest seaport would have been somewhere in Oklahoma or Tunisia as no Atlantic Ocean or Gulf of Mexico then existed. But large continental assemblages such as Pangaea are inherently unstable because the heat produced by radioactive decay in the continental crust and upper mantle keeps the Earth stirring.

Toward the end of the Triassic Period, masses of hot rock rising within the Earth's mantle began to rend Pangaea asunder along a series of rifts much like the modern Rift Valleys of Africa or the Dead Sea of the Middle East. One series of rifts connected together to become the incipient Atlantic Ocean, at one time perhaps much like the modern Red Sea. Other rifts never made the big time in the ocean business, but continued to grow and deepen as the continental crust "necked down" to meet the new oceanic crust at the base of the continental rise.

These rifts mostly hinged down from their eastern sides and received sediments mostly from their west sides; they are half grabens. Filling and deepening continued apace with the earliest sediments acquiring the steepest dips as the basin continued to open. Analysis of the sediments shows that some accumulated in standing water; others were laid down by running water of streams; still others, such as the carbonate pebble conglomerate of the Culpeper basin, were debris flows. The Farmville, Richmond, and Taylorsville basins supported swamps that later became coal seams.

The northernmost of the basins, the Culpeper, was intruded by diabase dikes, which periodically fed basalt lava flows that covered the surface. The other basins lack recognized lava flows, but do contain dikes of diabase. Beyond the borders of these basins, other dikes of Jurassic age intrude rocks of the Piedmont geologic province. Many of these dikes are large enough to have fed surface flows that have since eroded away.

The southernmost of these basins in Virginia, the Danville basin, is associated with the Swanson uranium ore deposit, largest yet discovered in eastern North America. Careful analysis of this deposit reveals that the ore is associated with the fault zone on the west side of the basin, not the Danville basin itself.

Wells show that the Taylorsville basin extends for 60 miles under the sediments of the Coastal Plain geologic province. Wells also reveal the existence of another Triassic basin 30 miles long buried beneath Coastal Plain sediments northeast of Richmond. Others, yet undetected, may exist beyond the shore line on the continental shelf. Results of geophysical studies indicate the existence of Triassic basins under Byrd Field at Sandston and under West Point on the York River.

Deepening and filling of the Triassic basins continued into Jurassic times. Small amounts of coal have been mined from the Richmond and Danville basins, but these deposits became unprofitable after discovery of the better coal fields of western Virginia and of West Virginia. The Triassic basins are also a target for oil and gas exploration.

All the basins in the Piedmont and under the sediments of the Coastal Plain were filled by the end of Jurassic time. Only the Triassic rift that became the Atlantic made it in the ocean business.

WHO'S PUSHING THE PLATES AROUND?

The lithospheric plates "float" on the asthenosphere, a zone in the upper mantle where seismic waves slow down due, presumably, to partial melting of the mantle rocks there. These plates are capped by light-weight, thick granitic continental crust and by heavy-weight, thin basaltic oceanic crust. Oceanic crust can be subducted—pushed under adjacent crust—but continental crust cannot. Continental crust has a larger content of heat-producing radioactive elements than oceanic crust. These statements are not theory; they are facts derived from detailed geological, geochemical, and geophysical measurements.

The theoretical shape of the Earth is the geoid—perfect mean sea level world wide. Gravity studies reveal that the geoid has humps and depressions on it. Continents and active subduction zones create humps; ocean basins over recently subducted cold oceanic crust are depressions.

Since the lithospheric plates "float" on the asthenosphere, continents will respond to gravity by tending to migrate toward any topographically low areas of the geoid. Once a continent arrives over one of these cold, low areas it has no further tendency to move on its own—unless bumped by a late-arriving continent, microcontinent, or island arc.

But wait. Recall that most of the radioactive heat sources are in the continental crust, with a little in the oceanic crust and upper mantle and almost none in the lower mantle. The thick, hot continents block heat flow coming up from the upper and middle mantle, causing subcontinental areas to heat and swell.

This heat can also cause partial melting in the upper and lower mantle, the major source of basaltic magma. Basaltic magma tends to rise through the mantle, but stops at the base of the crust, where it has temporary neutral buoyancy. During that pause, basaltic magma cools and gets lighter. Dense minerals crystallize out, as less dense material dissolves in from the surrounding host rock. Because the heat is sufficient to melt overlying continental crustal rock, the source for granites and rhyolites is often associated with basaltic magmatism.

So the continents are the source of their own destructive energy. Rising of the continents makes them tend to move downslope on the geoid, so the continents split and begin to move again, with fresh basaltic magma rising to fill the new rifts. The continents then migrate to new lows in the geoid subducting old oceanic crust beneath their bows while leaving a trail of new oceanic crust in their wake.

A special instability occurred in Paleozoic time when Gondwana assembled on a polar region about 400 million years ago. Geologists record large latitude shifts between then and the break up of Pangaea at the end of the Paleozoic Era.

Crustal plate movement is episodic. First the continents find cool, low spots and the whole crust shifts. Then the continents heat up the mantle, get uncomfortable, and move on. This extends as far back in geologic time as geologists can see. Gravity and radioactive heat move the plates around.

Quantico slate with sand lenses crops out in the Interstate 95 south exit to Quantico.

US 1, Interstate 95
Alexandria — Fredericksburg
48 mi./77 km.

These routes enter Virginia at the Potomac River on Cretaceous clays, silts, sands, and gravels of the Coastal Plain. These flat-lying strata lap onto the Precambrian and Paleozoic igneous and metamorphic rocks of the Piedmont. The boundary between the two geologic provinces is the Fall Line, so called because of waterfalls and rapids in stream channels where they cross it.

The routes parallel the Fall Line on the Coastal Plain side for most of the distance between Alexandria and Fredericksburg. Some of the streams have cut through the soft sedimentary rocks of the Coastal Plain into the basement rocks beneath. At Occoquan, the Occoquan River eroded through the Cretaceous sediments to expose the Occoquan granite and the Quantico slate. At Quantico, the southbound exit from Interstate 95 is cut into slates and sandstones of the Quantico formation. At Fredericksburg, the Rappahannock River and its tributary, Falls Run, cut their channels into the Po River gneisses.

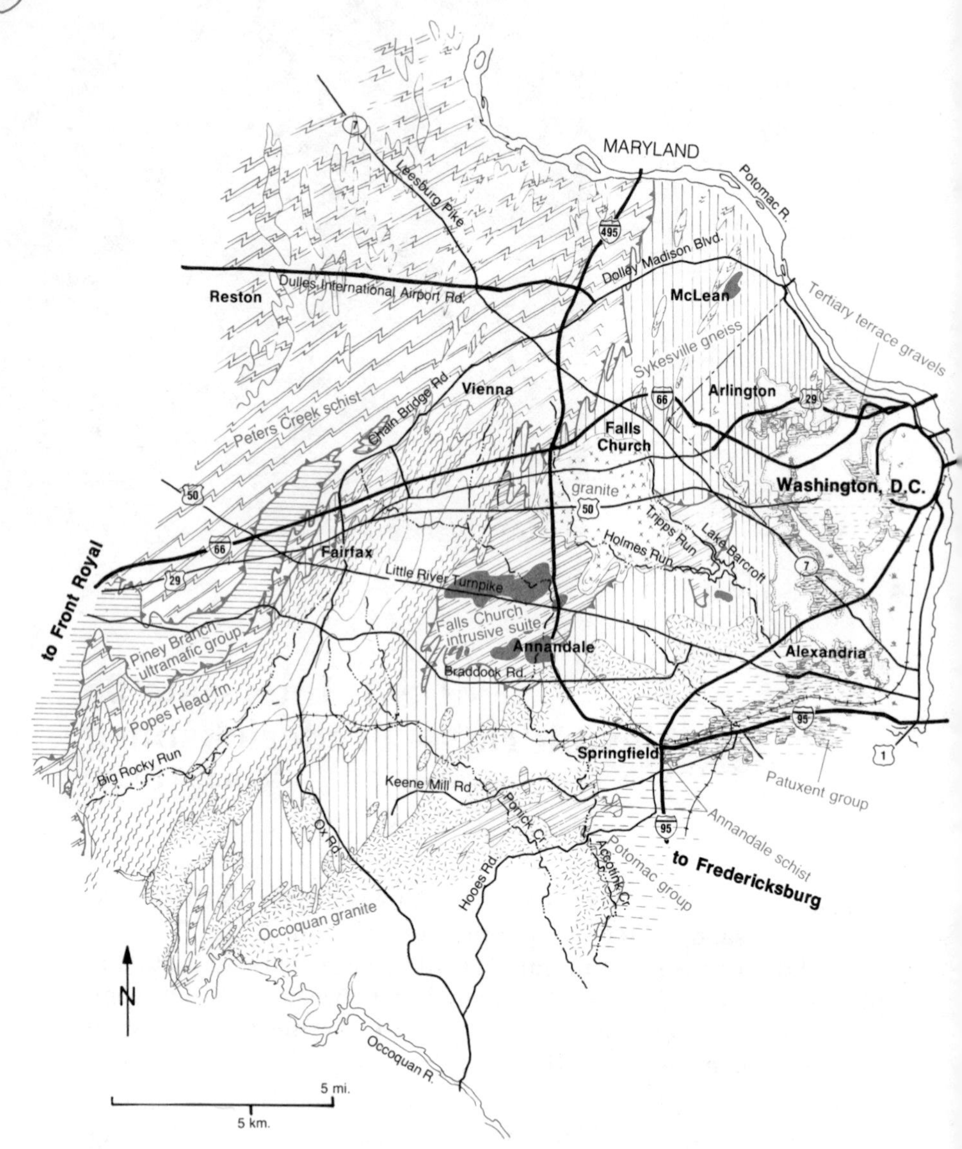

ALONG THE POTOMAC RIVER

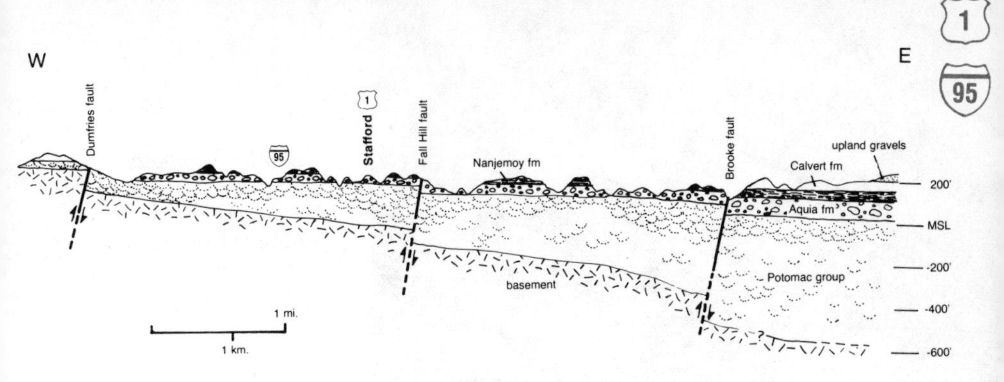

Reverse faults of the Tertiary break the Coastal Plain sediments between Alexandria and Fredericksburg. These faults cause the rock formations to step down toward the Atlantic Ocean.

The routes cross the Fall Hill fault, US 1 north of Falmouth and Interstate 95 at the Virginia 3 interchange. Geologists now think this fault is an extension of the Spotsylvania lineament, a major boundary dividing the Piedmont from here to the North Carolina state line. Although the lineament is probably Paleozoic in age, it was later reactivated to produce this and other faults in the Coastal Plain.

Exotic fragments of all sizes are common in the Sykesville formation. Ramesh Venkatakrishnan photo.

US 29
Arlington — Culpeper
67 mi./108 km.

From the south end of the Key Bridge, the road crosses from the Coastal Plain onto Precambrian igneous and metamorphic rocks of the Fredericksburg complex. Away from the Potomac River Valley, bedrock is hidden by deep saprolite and young river terrace deposits.

Centreville stands on a basalt lava flow near the eastern margin of the Culpeper basin, which filled with sediments and basalt flows during Triassic and Jurassic time. Between Centreville and Bull Run, also on basalt, the route crosses sedimentary basin fill. Except for topographic highs underlain by basalt, the surface of the basin is extremely flat, even when compared with the surface carved on the Fredericksburg complex. At New Baltimore, notice the sharp change in landscape between the Culpeper basin to the east and the more rugged ground underlain by more resistant Catoctin greenstone to the west.

Warrenton is on the Catoctin greenstone that makes up the eastern limb of the complex fold of the Blue Ridge anticlinorium. The Blue Ridge proper is the western limb, with crystalline rocks of the Precambrian basement cropping out between the two greenstone belts. South of Warrenton, where US 17 joins from the south, the route crosses the western border fault back onto the flat topography of the Culpeper basin. The route continues on a loop into the basin north of Culpeper, which is built right on the western border fault. Again, notice the change in the landscape on opposite sides of the fault.

Virginia 55, Interstate 66
Arlington — Strasburg
74 mi./120 km.

West of the Potomac River valley the route leaves the Coastal Plain and crosses onto the igneous and metamorphic rocks of the Precambrian Fredericksburg complex. The bedrock is covered with saprolite and with river terrace deposits, so you see no outcrops.

Centreville is on the eastern side of the Culpeper basin. Bedrock there is one of the basalt flows that helped fill the rift basin as it formed during Triassic and Jurassic time, while the Atlantic Ocean was opening. Weather-resistant basalt dikes support the few ridges in this otherwise monotonously flat basin.

The routes cross the western border fault of the Culpeper basin just east of Thoroughfare Gap and climb onto Catoctin greenstone. Note the dramatic difference in topography between the rugged land underlain by greenstone and the flat ground in the basin to the east.

Between The Plains and Markham, the routes run across the basement complex in the core of the broadly folded arch of the Blue Ridge anticlinorium. The Blue Ridge is held up by Catoctin greenstone, metamorphosed basalts of latest Eocambrian age. Watch for outcrops of these greenstones in the roadcuts at the crest of the Blue Ridge.

West of the Blue Ridge, the route descends through the early Cambrian Chilhowee sandstones to the Shenandoah Valley. Rocks there are formations of the early Paleozoic carbonate bank. An upedged section of this bank lies between the foot of the Blue Ridge and the center of Front Royal. Between Front Royal and Strasburg, the routes traverse the complexly folded trough of the Massanutten synclinorium on shales and sandstones of the Martinsburg formation.

From Strasburg, Little North and Great North mountains are visible to the west, Massanutten Mountain to the south, and the Blue Ridge to the east. Between Little North and Massanutten mountains, the valley of the North Fork of the Shenandoah River stretches off to the southwest.

Massanutten Mountain occupies the central portion of the Massanutten synclinorium and is carved out of the Massanutten sandstone. The folded synclinorium trough extends both north and south of the mountain where it is underlain by the Martinsburg shale. Waterlick is in the trough of the fold. The road south from Waterlick leads into the valley eroded right along the trough.

Anticline and syncline in thick limestone bed in the early Paleozoic carbonate bank. Photo by Thomas M. Gathright, II; Courtesy of the Virginia Division of Mineral Resources.

Pieces of Marshall granite and vein quartz picked out of fields were used in construction as in this wall of an outbuilding to a former stagecoach inn near Middleburg.

US 50
Arlington — West Virginia state line
79 mi./127 km.

The route begins in Cretaceous sedimentary rock of the Coastal Plain. Above the river, the route runs onto igneous and metamorphic rocks of the Piedmont, the Precambrian Fredericksburg complex. The bedrock is buried beneath a thick mantle of deeply weathered saprolite, in some places topped by stream gravels. The western edge of this crystalline complex is in Greenbriar.

The route crosses the Culpeper basin between Greenbriar and Gilberts Corner. These elongate basins opened as the supercontinent split when the Atlantic Ocean basin began to open during late Triassic to early Jurassic time, some 180 million years ago. The sharp rises in the highway between Aldie and the junction with US 15 mark weather-resistant diabase dikes that the roadway drapes across. They also date from the opening of the Atlantic Ocean. The western fault boundary of the basin surfaces just west of these dike ridges.

A narrow strip of metamorphic rock separates the late Precambrian Catoctin greenstone from the western border fault. Catoctin greenstone holds up Catoctin and Bull Run mountains and US 50 follows the pass between them. Note the strong contrast between the rugged topography eroded in the weather-resistant greenstone and the much flatter landscape eroded into the flat-lying sediments and basalt flows of the Culpeper basin to the east.

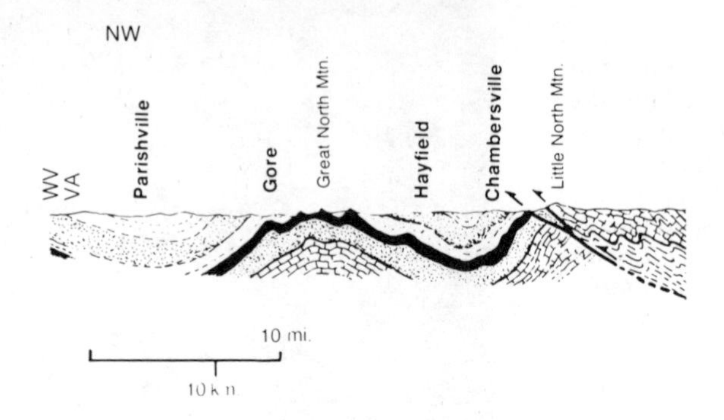

Cross section of the Valley and Ridge geologic province

Sheared granite appears in the roadcut on the west side of Middleburg. Walls, foundations, and many of the old homes are made of it. The variety of sizes and shapes of the individual stones seems to indicate that builders opportunistically collected appropriately shaped stones.

Pavements of coarse grained diorite, a dark igneous rock, crop out in the fields near Upperville. The variety of local rocks can best be studied in the stone walls that line both primary and secondary roads in this area. Local legend relates that these walls are made of stone cleared from the fields by unemployed Hessian soldiers after the American Revolution.

About three miles east of Ashby Gap the route crosses from the basement crystalline complex onto younger Catoctin greenstone. Greenstones hold up Catoctin and Bull Run mountains to the east. Ashby Gap in the Blue Ridge is cut into greenstones, metamorphosed basalt flows and volcanic ash falls of latest Precambrian age. Greenstone is the dominant ridge former in the northern Blue Ridge.

Between Ashby Gap and the Shenandoah River, the rocks get younger as you travel down into the valley, and vice versa. US 50 crosses the Shenandoah River where it cut its channel into the youngest formation of the Chilhowee group. Between the Shenandoah River and Opequon Creek, the route crosses Cambrian to Ordovician limestones of the early Paleozoic carbonate bank. Limestone is abundant in the fields on either side of the highway and in the many roadcuts. The rolling topography is typical of the limestone portion of the entire Shenandoah Valley.

Between Opequon Creek and Winchester, the route crosses Or-

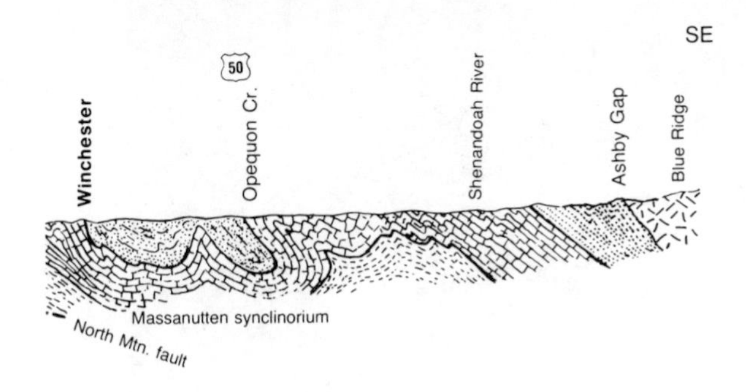

parallel to US 50 between West Virginia and the Blue Ridge.

dovician sedimentary rocks and the axis of the Massanutten synclinorium. The resistant Massanutten sandstone is missing here, so the fold has no strong topographic expression. Between Winchester and Little North Mountain the route crosses Ordovician and Cambrian sedimentary rocks on the west flank of the synclinorium.

Little North Mountain marks the midprovince structural front where Cambrian sedimentary rocks of the early Paleozoic carbonate bank were shoved onto younger Silurian formations along the North Mountain thrust fault. Chambersville is on the slice between two branches of the fault. In the gap through Little North Mountain red sandstone of the Silurian Clinton formation is exposed in the roadcuts.

Great North Mountain is a broad folded arch, an anticlinorium, cored by Tuscarora sandstone. Watch for loose blocks of that sandstone next to the eastbound lane where the route wraps around the nose of the arch. Between Great North Mountain and Little North Mountain, visible to the east, the rocks are folded into a synclinal trough.

Near the West Virginia state line you look down into a valley eroded into Devonian shale buckled down into a syncline. The first ridge to the east is Little Timber Mountain, the opposite side of the syncline. Great North Mountain lies beyond. New roadcuts west of Gore reveal near vertically dipping beds on the east side of the syncline.

Folds in Devonian sandstones and shales exposed in roadcuts east of the West Virginia state line.

Virginia 7
Alexandria — Winchester
70 mi./113 km.

From Alexandria, the route crosses the Fredericksburg igneous and metamorphic complex. Those rocks are well hidden beneath a deeply weathered layer of saprolite covered by terrace sand and gravel deposits. The eastern margin of the Culpeper basin is in Dranesville. Several faults at this margin raised a block of Peters Creek schist into the Triassic rocks and dropped a block of Triassic rocks into the schist. About a mile west of Dranesville, Triassic sandstone and siltstone crop out on the north side of the road.

About 500 yards south of the junction with Virginia 28, diabase crops out in the roadcut. Where the diabase is homogeneous and relatively free of fractures, it is quarried for dimension stone and sold as black "granite." Where it is much fractured and more variable in texture, it is crushed into tough, weather-resistant road aggregate. About two miles east of the Leesburg bypass interchange the road crosses Goose Creek. One quarter mile farther west, the westbound lane is cut into light to dark olive-colored hornfels produced by contact metamorphism of the sedimentary rock of the basin by heat from the intrusive diabase. These diabase dikes form the prominent ridges that project above the generally flat surface eroded on the sediments. Leesburg is on the contact between diabase and carbonate pebble conglomerate that runs parallel to Virginia 7 to the interchange east of town. Good exposures of the conglomerate appear along business US 15 just north of Leesburg.

The highway crosses the fault on the west side of the Culpeper basin near the Virginia 7 bypass interchange west of Leesburg. Both Catoctin greenstone and sheared sandstone of the fault zone crop out in the off ramp for bypass Virginia 7. Note that the landscape is less rugged in these sediments than on the more resistant granitic rock on the west side of Catoctin Mountain.

The route crosses Catoctin Mountain at Clarks Gap where greenstone altered to a rusty red color crops out on the north side of the road. This eastern belt of greenstone extends from the bypass interchange to just west of Hamilton. Watch in fresh excavations for the red soil that develops on the greenstone.

Between Hamilton and Round Hill, the route crosses the granitic core of the broad arch of the Blue Ridge anticlinorium. The landscape eroded into these granites and granite gneisses, here called the Marshall formation, is gently rolling and affords few outcrops. The tan and sandy soil you see in fresh excavations contrasts sharply with heavy red soil on the flanking greenstone belts. The Blue Ridge is carved out of Catoctin greenstone of latest Precambrian age. On the east side of the Blue Ridge the view is all the way across the Blue Ridge anticlinorium to Catoctin Mountain.

The Chilhowee sandstone forms a ridge of its own just west of the Blue Ridge with a good cut in it on the south side of the road. The view west from the ridge is a nice panorama of the Shenandoah Valley. The Shenandoah River here cuts its channel into the Chilhowee sandstones, although most of its course southwest of here is in the early Paleozoic carbonate bank. The route climbs the geologic column into the carbonate bank between the Shenandoah River and the east bluff of Opequon Creek. Outcrops of steeply dipping carbonate rock in the many fields west of Berryville tend to be in the briar patches that farmers have not mowed for fear of damage to their equipment on a rock.

The Martinsburg shale between Winchester and Opequon Creek is topographically more dissected than the carbonate bedrock on either side of the synclinorium. Good exposures of the Martinsburg formation are south of the eastbound lane. The axis of the Massanutten synclinorium is approximately two miles east of the city limits of Winchester. The north end of Massanutten Mountain lies on the trend of this fold, some 18 miles to the southwest.

Winchester is on rocks of the early Paleozoic carbonate bank on the west flank of the Massanutten synclinorium. Local limestone is used in retaining walls, foundations, and the old railroad station. The boundary between the carbonate rocks and Ordovician sedimentary rocks lies just west of the interchange with Interstate 81 on the east side of town.

This carbonate pebble conglomerate slumped into a lake that once occupied the Culpeper basin.

US 15
Potomac River — Culpeper
66 mi./106 km.

The route enters the state at the narrows the Potomac River created where it cut through resistant greenstone of Catoctin Mountain. The first roadcuts are in sheared Catoctin greenstone dappled with numerous white quartz lenses, a metamorphosed basalt flow of latest Precambrian age. As the route climbs out of the valley, it swings slightly eastward across a narrow strip of early Cambrian Chilhowee sandstones, then crosses the western border fault of the Culpeper basin. The basin is a block that filled with sediments as it dropped along faults while the Atlantic Ocean began to open during Triassic and Jurassic time.

Bedrock in this part of the Culpeper basin is the carbonate pebble conglomerate used for ornamental "marble" in the District of Columbia and elsewhere. This rock forms blocky and pavement outcrops in the fields on both sides of the road. Many roadcuts expose it between the river and Leesburg; some of the best exposures are on business US 15 south of the bypass intersection. Solution of the limestone by ground water creates a very irregular topography with numerous ponds and sinkholes.

Leesburg is at the intersection of US 15 with Leesburg Pike, now Virginia 7, once the main route from Washington to Winchester and to the spa at Berkeley Springs, West Virginia. The intersection is near the contact between a diabase intrusion and baked sedimentary rocks nearby. The low ridge on the east side of the highway about a mile south of the bypass interchange is held up by a diabase dike.

Between Sycolin Creek and Oatlands, roadcuts of sheared rock mark the western border fault of the basin. Jurassic bedrock immediately east of the fault is covered with debris shed off Catoctin Mountain. There is a good exposure of Jurassic basalt with a fault offset in Goose Creek. At the junction with US 50, Bull Run Mountains, the southern continuation of Catoctin Mountain, has swung west from the highway. The low ridges between the road and the intervening low, but sharp, ridges are resistant Jurassic basalt. Gaps in Bull Run Mountain open to glimpses of the Blue Ridge in the distance.

On both sides of the junctions with Virginia 55 and Interstate 66 the basin-fill consists of thin layers of limestone and siltstone deposited in a lake. At Haymarket, just south of the railroad tracks on the east side of the road, a basalt lava flow overlies red silty sandstone on the south side of a parking lot. These rocks are of earliest Jurassic age. The basalt flow has closely spaced joints at right angles to the gently dipping beds.

South of the junctions with Virginia 55 and Interstate 66 the ridge held up by Catoctin basalts on the west flank of the basin is called Pond Mountain. About a mile south of the junction with US 29, the southbound lane is cut into another basalt flow. Basalt crops out

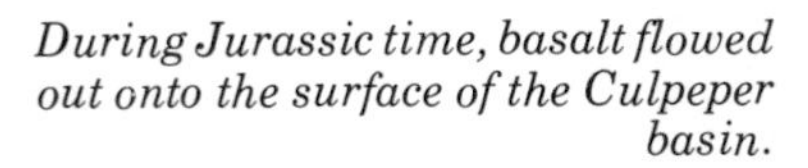

During Jurassic time, basalt flowed out onto the surface of the Culpeper basin.

Slickensides gouged shear along a fault in the Catoctin greenstone are decorated with white quartz and green epidote.

along both lanes around the junction with highway 215. The route crosses out of the Culpeper basin near the village of New Baltimore. Note the distinct change in topography between the fairly flat surface of the basin and the rolling terrain underlain by the more resistant greenstone. Warrenton is on Catoctin greenstone.

South of Warrenton, US 15 again drops off the greenstone back into the Culpeper basin at about the intersection with highway 600. Here the bedrock is basalt, but the route crosses onto basin sediments about half way to the town of Remington. Outcrops and roadcuts are sparse. Culpeper lies on the western border fault of the Culpeper basin.

Skyline Drive
Shenandoah National Park
Front Royal — Swift Run Gap
66 mi./106 km.

Skyline Drive winds back and forth across the crest of the Blue Ridge for 105 miles, affording views to the east and west from its many overlooks. If you enter Shenandoah National Park at the south end, use the park map to follow this section from back to front.

Three main rocks crop out in roadcuts and cliffs along the route—greenstone, granite, and sandstone. *Federal law forbids rock or mineral collecting within the park.*

Greenstone is basalt that has been transformed by heat and pressure during moderately deep burial in the Earth's crust. When later exhumed by erosion, these metamorphosed rocks crop out as massive formations. Interbedded with basalt flows, however, are volcanic ash falls, the products of more explosive eruptions. Metamorphism transforms these into slaty rocks which, in the Blue Ridge, are also called greenstone.

Granite, in the strict sense used by geologists, is a rock formed from molten magma deep in the crust of the Earth and composed of about one third quartz and two thirds feldspars. Generally, less than ten percent of the rock is white or black mica. In the park, there are rocks which are, like strict granite, predominantly feldspar and quartz, but also contain the dark-colored minerals garnet and pyroxene. The name for this type of granite is charnockite. In addition, there are banded rocks, strictly speaking gneisses, associated with the Pedlar formation. Granite as used here includes all these feldspathic rocks.

Sandstone is a rock composed of quartz sand grains and quartzite is a sandstone that has been metamorphosed by the heat and pressure of burial in the crust of the Earth. Grains coarser than sand make the rock a conglomerate and those finer, a siltstone. If feldspar sand grains are in the rock, it is an arkose. Although clastic rock is the catch-all name for these rocks, sandstone is a perfectly good general term. In the park, however, many geologists use sandstone and quartzite interchangeably.

From the north entrance to the park south of Front Royal to Lands Run Gap, Skyline Drive is in Catoctin greenstone. Where fresh and unaltered by rain and ground water, this rock is dark gray-green to black. On outcrop and freshly exposed joints in roadcuts, it is typically a medium gray-green although the roadcuts near the entrance are stained red by iron oxide.

Overlooks to the west provide a panorama of the South Fork of the Shenandoah River, Massanutten Mountain, and in the far distance on very clear days Little North Mountain and Great North Mountain. Here the West Virginia state line runs along the rear flank of Little North Mountain. The Shenandoah Valley extends from the west slopes of the Blue Ridge to the east slopes of the Little North Mountain, neatly bisected by Massanutten Mountain for nearly 50 miles.

The nearly level crest of Massanutten Mountain is held up by Silurian Massanutten sandstone and much of the visible lower slope is mantled by loose talus blocks of sandstone. The South Fork of the Shenandoah River meanders across a surface underlain by Cambrian and Ordovician rocks, mostly limestones and dolomites, of the early Paleozoic carbonate bank. Much of that river's course appears to follow fractures opened by solution of the carbonate bedrock.

Gooney Manor Overlook faces nearly south across Gooney Run and its tributaries. Bedrock in this watershed is Pedlar granite, but the Blue Ridge on the far side is capped by Catoctin greenstone.

In roadcuts between Lands Run Gap and Indian Run Overlook, nearly vertical dikes of diabase are exposed. Geologists think that these dikes fill the fractures that carried basalt from magma chambers beneath the Earth's crust to the Precambrian surface, where the magma erupted to become lava flows of the Catoctin formation. Only those dikes thicker than 25 feet, however, could have contributed to any substantial lava flow.

At many locations in the park, the basalt flows that became the Catoctin greenstone spread themselves directly on the underlying granite. At others, such as at Indian Run Overlook, the Swift Run formation lies between the top of the granite and the lowest basalt flow. Here the Swift Run is a medium to coarse grained sandstone. Some of the greenstone at this location shows distinct columnar jointing that makes the exposure look like a log stockade.

Eastward-facing overlooks reveal the core of the Blue Ridge province, distant ridges held up by the Lovingston and Lynchburg gneisses and more Catoctin greenstone. Bedrock under the eastern slopes of the Blue Ridge here is granite like that at Jenkins Gap and Hogwallow Flat overlooks.

The contrasting styles of the ridges and mountains on either side of the Blue Ridge show clearly from Range View Overlook. Upturned resistant sandstone beds support the ridges of the Valley and Ridge province to the west. The result is long, even-crested ridges such as Massanutten Mountain. Ridges to the east are eroded in massive granite, gneiss, and related rock cut by faults and shear zones, some parallel to the Blue Ridge, others cutting across the trend of the ridges and valleys. Erosion of sheared rock creates discontinuous strings of hills with various elevations.

At Range View Overlook, veins and pods of the apple-green mineral epidote and white quartz cut the Catoctin greenstone. In the roadcut across from the overlook, pods of sandstone indicate that the Catoctin formation contains water-deposited units as well as basalt flows.

The contact between the granite and the greenstone lies downslope from Gimlet Ridge Overlook and crosses Skyline Drive just east of Little Hogback Overlook. To enjoy some of the difficulties of working out the details of Blue Ridge geology, find the closest distance between outcrops of granite and of greenstone and try to imagine some of the different kinds of things geologic that might be hidden in that covered interval.

Diabase dikes cut the Pedlar granite at milepost 20. The granite at Little Devil Stairs Overlook is a rather typical charnockite with fractured crystals of feldspar the size of a thumb or larger set in a matrix of smaller feldspar, quartz, and dark-colored minerals that were originally pyroxene or garnet. Because the contact with the overlying Catoctin greenstones is so close, some of the red iron-oxide stain in this granite may be the result of Precambrian rock weathering that happened before eruption of the Catoctin basalts.

Both north and south of Jeremys Run Overlook, the Blue Ridge is capped by Cambrian pebble conglomerates, slates, and sandstones of the Chilhowee group. Knob Mountain has the same caprock, but the

Outcrops of Pedlar charnockite are typically rounded as at Little Devil Stairs Overlook on Skyline Drive.

lower slopes above Jeremys Run are cut into greenstone. The notch in Massanutten Mountain is New Market Gap where US 211 connects Luray with New Market.

Thornton Gap formed through differential erosion of crushed rock along the Stanley fault, which crosses the Blue Ridge here. This fault runs from Stanley on US 340 in the Shenandoah Valley to US 522 about 7 miles southwest of Front Royal east of the Blue Ridge. Throughout most of its length, Pedlar granite was shoved from the southeast over Catoctin greenstone along the fault. The first outcrop north of US 211 is greenstone and the first one south is granite.

From Thornton Gap to Stony Man Overlook, Skyline Drive travels on Pedlar granite. Most of the mountains to the east—Thorofare, Corbin, Robertson, and Old Rag itself—are carved out of the Old Rag granite. Thumb-sized and larger feldspar crystal in a matrix of blue quartz and smaller feldspar grains characterize this granite, which typically carries very little dark-colored mineral. Much of the quartz is severely fractured, causing it to weather out first to leave the large feldspar crystals standing out like knobs on one side of a giant waffle press.

Stony Man resembles a human face carved by erosion from columnar basalt in the Catoctin formation on the Blue Ridge.

Stony Man Overlook provides a view of Luray. The sharp, tree-covered ridges in the near distance are cut from sandstones of the Cambrian Chilhowee group. The rolling surface of the Shenandoah Valley is on limestones of the early Paleozoic carbonate bank. Ground water first dissolved caverns in those carbonates and then deposited all the spectacular underground rock forms that make some of these caverns commercial successes.

The crest of Massanutten Mountain is the Silurian Massanutten sandstone, talus blocks of which cover much of the lower slope. The notch in the crest of the mountain ridge, New Market Gap, is where erosion cut all the way through the resistant sandstone into the more easily eroded Ordovician Martinsburg formation, here a sandstone.

Stony Man Mountain, so named because some people imagine they can see a human face outlined against the sky along its western slopes, is supported by Catoctin greenstones. The contact between the granite here and the greenstone there is near the Stony Man Trail parking area a few hundred yards to the south. Walk up the trail a few yards, then take the right fork for about 25 yards. Outcrops of coarse-grained granite and fine-grained greenstone stand within a couple of yards of each other between the trail and Skyline Drive.

Between Skyland and Big Meadows, Skyline Drive crosses and recrosses the contact between the Pedlar granite and the Catoctin greenstone. The Swift Run sandstones and mudstones separate the two formations for about two thirds of this distance, but are absent for the remainder. Folds and fault slices in the Chilhowee sandstones underlie the sharp ridges and precipitous slopes between the Blue Ridge and the Shenandoah Valley to the west. These ridges contrast strikingly with the even, parallel ridges of Massanutten Mountain and the knobby granite mountains and hills to the east.

The view from Black Rock, just west of Big Meadows Lodge, is one of the most spectacular along the entire Skyline Drive. Much of the length of Massanutten Mountain is visible and, on clear days, Little North Mountain beyond. US 340 swings west from Stanley around the rugged hills of complexly folded Chilhowee sandstones. Manganese was mined from these rocks for many years, especially during the world wars. The complex structural front of the Blue Ridge stretches away to the north.

The south end of Massanutten Mountain comes into view from the Oaks Overlook. Although the mountain ends, the synclinal structure continues on to the southwest in the less resistant Martinsburg shale. The roadcut 100 yards north from the overlook is a typically massive basalt flow of the Catoctin greenstone. Above that cut is a gently sloping covered interval before another greenstone cliff rears up in the trees.

The roadcut at Hensley Hollow Overlook reveals pebbles and cobbles in a conglomeratic portion of the Swift Run formation. The contact with the Pedlar granite is just below the overlook and the contact with the Catoctin greeenstone is just above the roadcut.

The Swift Run formation is named for outcrops east of Swift Run Gap in the woods above US 33. This gap differs from Thornton Gap in that no fault or shear zone has been identified to account for its low elevation relative to the rest of the Blue Ridge.

Catoctin greenstone typically forms jagged outcrops as at Oaks Overlook on Skyline Drive.

US 340
Maryland state line — Elkton
78 mi./126 km.

This route follows the trend of the early Paleozoic carbonate bank along the South Fork of the Shenandoah River. Limestone crops out in the fields between the Maryland state line and Front Royal. It has been used in fences, retaining walls, and many of the old buildings in Berryville and Front Royal. From Front Royal to Elkton the valley is divided by Massanutten Mountain. The complex downfold of the Massanutten synclinorium, however, extends from the Maryland state line to Fairfield, 13 miles beyond the southern terminus of the highway. Only in its central portion does the resistant Silurian Massanutten sandstone make the synclinorium a part of the landscape.

East of US 340 is the Blue Ridge, the westernmost ridge of the Blue Ridge province. For much of its length parallel to this part of the valley, the Blue Ridge is underlain by Grenville Pedlar granite and related rock and capped by Catoctin greenstone, basalt flows and related rocks deposited on the Pedlar granite in latest Precambrian time.

Many precipitous hills and ridges between the Blue Ridge and the Shenandoah Valley dominate views to the east of US 340. These hills and ridges were carved in folded fault slices of siltstone and sandstone of the Chilhowee group of rock formations.

Massanutten Mountain to the west is held up by the Silurian Massanutten sandstone here folded into a syncline or, more accu-

Calcite precipitates on a cavern ceiling. Photo courtesy of Skyline Caverns.

rately, a synclinorium, because the big fold has smaller folds superimposed on it. The upward tilted edges of this sandstone support a double mountain, a pair of parallel ridges. From some overviews on Skyline Drive, both are visible. A distinct valley lies in the fold between the two ridges in the northern half of the mountain. The Ordovician Martinsburg shale lies beneath the resistant, ridge forming sandstone. Erosion of the Martinsburg formation undercuts the Massanutten sandstone, causing blocks of it to break off and slide down to mantle many of the lower slopes of the mountain with talus.

The Shenandoah Valley on either side of Massanutten Mountain is carved into limestone and dolomite formations of the early Paleozoic carbonate bank. Locally prominent ridges and hills within the valley are supported by resistant zones of sandstone and chert.

Limestone and dolomite are carbonate rocks and, hence, susceptible to solution and removal underground by water percolating along bedding planes and fractures, especially where two such surfaces intersect in a line. This process leads to cavern formation, cavern collapse, and the development of a sinkhole dominated landscape known to geologists as karst topography.

From Front Royal to Luray, Chilhowee rocks are thrust over the carbonates of the Shenandoah Valley along the Blue Ridge front. In Browntown Valley, however, the tributaries to Gooney Run have eroded away all the Catoctin and Chilhowee rocks, leaving low hills cut into the Pedlar granites and related rocks. The thrust fault between the Precambrian granulite gneisses of the Pedlar massif of the Blue Ridge province and the underlying Ordovician carbonate rock of the Valley and Ridge province surfaces on the south side of Gooney Creek campground. This fault caused the underlying carbonate rocks to be overridden and overturned to the northwest.

South of Bentonville, the sharp hills and ridges just east of the highway are eroded into the Chilhowee sandstone. Knob Mountain, east of Riley, is an anticlinal arch overturned to the northwest with Chilhowee sandstone on its crest and lower slopes, Catoctin greenstones on the middle slopes.

Near Luray, the front of the Blue Ridge swings back to the east permitting local widening of the Shenandoah Valley. To the west, you see New Market Gap as a notch in the crest of Massanutten Mountain. Here the syncline is arched up and erosion has removed the protective sandstone cap from the mountain to expose the less resistant Martinsburg sandstones in the gap.

From Luray to Stanley, the highway parallels Hawksbill Creek in its course across the carbonates of the valley. The sharp ridges and hills to the east are more folds and fault slices of Chilhowee sandstone. In the distance is the Blue Ridge with western slopes cut in Pedlar granite and peaks of Catoctin greenstone.

South of Stanley, the highway swings west for about 5 miles to go around a bulge in the Blue Ridge front and crosses over the South Fork of the Shenandoah River. Between Stanley and Shenandoah in the Chilhowee sandstone hills to the east, manganese was mined sporadically from a dozen different pits for about a century, especially during the two world wars. Another use for these sandstones is revealed in a mountain name east of Shenandoah, Grindstone Mountain.

From Shenandoah to Elkton, the highway is on terraces and other alluvial material washed off the Blue Ridge to the east. You can sample the different kinds of rocks from the Blue Ridge in many of the streams which drain off it into the South Fork of the Shenandoah River.

US 522
West Virginia state line — Culpeper
89 mi./143 km.

The route enters the state in Sleepy Creek Valley, which is cut into Devonian shales. To the west is Cacapon Mountain, an anticline cored by Tuscarora sandstone. The white debris slides are blocks of sandstone shed off bedding planes in the formation and are typical of Tuscarora ridges throughout the state. This sandstone is quarried in West Virginia for high quality quartz sand.

Between the state line and Gainesboro, the route diagonally crosses a syncline in Devonian shales. At Gainesboro, the highway wraps around the end of Great North Mountain, another anticline cored by Tuscarora sandstone. This structure heaves into view about six miles south of the state line. Most of the roadcuts along this stretch of highway are in the Devonian formations. Hunting Ridge west of Hogue Creek is a syncline of Devonian shales and sandstones.

Between Sunnyside and Albin, Cambrian carbonate rocks were shoved onto younger Silurian and Devonian sandstones and mudstones along the North Mountain thrust fault at Little North Mountain. This fault is the western boundary of the Shenandoah Valley. From Little North Mountain west of Winchester to the Blue Ridge east of Front Royal, US 522 angles across the Shenandoah Valley and the complexly wrinkled downfold of the Massanutten synclinorium.

South of Winchester near the Interstate 81 interchange, the route passes out of the early Paleozoic carbonate bank on the west side of the synclinorium into the Ordovician Martinsburg shale. Near the bridge over Opequon Creek, the road crosses the axis of the synclinorium. Massanutten Mountain is clearly visible to the south between Winchester and Front Royal. From Armel to Front Royal the road follows the trend of the Martinsburg formation parallel to the carbonate rocks to the east. Just north of the bridge over the Shenandoah River, the highway passes through a cut in the rocks of the carbonate bank that reveals ripple marks in the bedding planes west of the southbound lane. Numerous white veins of calcite fill fractures in the rock.

Southeast of Front Royal, the route crosses the fault at the base of the Blue Ridge onto Catoctin greenstone, metamorphosed basalt, of Eocambrian age. Along fracture surfaces in the roadcuts on the outskirts of Front Royal the rock is weathered to a dull red color but fresh chips show the original olive-green color. The route cuts through the Blue Ridge at Chester Gap and onto the Precambrian crystalline rocks in the core of the Blue Ridge anticlinorium near Huntly.

From Huntly to the junction with US 211 at Massies Corner, the bedrock is the Flint Hill gneiss, a layered rock formed by metamorphism of quartz monzonite, an igneous rock fairly similar to granite. Large pavement outcrops of this formation exist on the low hills above this intersection. The dominant peak visible ahead from southeast of the intersection along combined US 522 and US 211 is Old Rag. The Old Rag granite is a distinctive unit poor in dark minerals. It contains pecan-sized crystals of feldspar, which weather out to make knobby outcrops. The fine-grained matrix consists of very small crystals of feldspar and blue quartz.

Between Massies Corner and "little" Washington, the bedrock is an

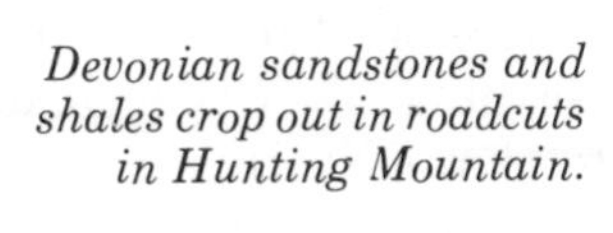

Devonian sandstones and shales crop out in roadcuts in Hunting Mountain.

Ripples on a shallow bank in the Iapetus Ocean preserved in the rock, and now exhumed in a roadcut along US 522 in Front Royal

augen gneiss of the Oventop suite. The augen are the thumb-sized crystals of pale feldspar. At "little" Washington the route is underlain by a layered gneiss that contains very few dark minerals and may correlate with the Flint Hill gneiss. At the western intersection of the "little" Washington bypass the gneiss is cut by a diabase dike, possibly a feeder to the Catoctin basalt flows. The outcrop at the intersection of US 522 and US 211 in Sperryville is either sheared Pedlar charnockite or part of the Flint Hill gneiss.

Southeast of Sperryville, the route crosses the Elkton fault, which here brings the Lovingston augen gneiss from the southeast onto Pedlar charnockite. From there to Woodville the bedrock is the Lovingston augen gneiss. Round Hill is in the middle of the two-mile wide Mechums River graben, a block dropped along faults that stretches from 10 miles north of this location to a point south of Interstate 64 west of Charlottesville. Sediments in the graben were laid down during the late Precambrian time, but before eruption of the Catoctin basalt lava flows. One mile east of Round Hill and a mile and a quarter west of Boston, the route crosses out of the graben and onto the Robertson River hornblende granite. Curiously, this granite is named for outcrops on the Robinson River, which is thusly misnamed on some maps.

Near Norman, the route climbs up the geologic section to the Lynchburg group of rock formations. These metamorphosed sedimentary rocks of late Precambrian but pre-Catoctin age are thought to be marine equivalents of the terrestrial sediments in the Mechums River graben. Between Norman and Culpeper, the route crosses two infolds of the overlying Catoctin greenstone. Culpeper sits on the western edge of the Culpeper basin, which formed as the Atlantic Ocean began to open during the Triassic, then filled with sedimentary rocks during Triassic and Jurassic time.

US 17
Winchester — Fredericksburg
80 mi./129 km.

The route is the same as US 50 from Winchester to Paris on the east side of the Blue Ridge. South of Paris, it crosses terrain eroded into the western belt of Catoctin greenstone. Between Interstate 66 and Old Tavern, the route crosses the crystalline core of the Blue Ridge geologic province. Here the bedrock is strongly sheared and fractured granitic rocks of the Marshall formation. South of Old Tavern, the road crosses the eastern belt of Catoctin greenstone. Note the change in the style of the topography to the south. Wherever it exists, greenstone erodes to distinctly rugged terrain.

South of Warrenton at Opal, where US 17 splits off US 15, the route crosses the western border fault of the Culpeper basin filled with sediments and basalts of Triassic and Jurassic age. Again note the change in the style of topography from that around Warrenton. The landscape within the Culpeper basin is notably flatter than the more rugged terrain on either side. Two miles north of Morrisville, the route leaves the Culpeper basin and crosses onto the Piedmont province.

Piedmont rocks along the route include igneous and metamorphosed sedimentary and volcanic rocks originally laid down in an early Paleozoic sea or chain of volcanic islands. At Goldvein, these metamorphic rocks were intruded by granitic rock, which was also metamorphosed. The town takes its name from the numerous gold-bearing quartz veins that were mined from the 1830s until the 1930s. The gold occurs with pyrite, fools' gold, which was processed into sulfuric acid. Mineral specimens still exist on the abandoned mine dumps, all on private property.

At Hartwood, the underlying metamorphosed volcanic and interbedded sedimentary rocks belong to the Chopawamsic formation, which extends south across the James River. Some geologists think the Chopawamsic and overlying formations constitute a separate terrane that was added to the Piedmont during Paleozoic time.

Southeast of Hartwood, the Chopawamsic formation is intruded by another metamorphosed granitic rock. South of the intersection with highway 616, rocks of the Fredericksburg igneous and metamorphic complex are covered by ancient terrace deposits. Upstream from the bridge over the Rappahannock River, the island is cut into gneiss of the Po River metamorphic suite. Little can be seen of bedrock in the Fredericksburg area because it was deeply weathered, then capped by terrace deposits.

Signal Knob marks the north end of Massanutten Mountain, but the Massanutten synclinorium extends northward into Pennsylvania.

US 11, Interstate 81
West Virginia state line — Harrisonburg
79 mi./127 km.

These routes parallel bedrock outcrops down the Shenandoah Valley, which is eroded into limestones and dolomites of the early Paleozoic carbonate bank formations. These rocks appear in roadcuts and fields along most of this stretch of road. Pumpkin Ridge is an anticline arched into limestones west of Winchester.

From Signal Knob east of Strasburg to just south of Harrisonburg, the Shenandoah Valley is sharply split by Massanutten Mountain. The major river in this part of the valley is the North Fork of the Shenandoah River, a tributary of the Potomac.

East of this part of the valley is Massanutten Mountain, a synclinal trough that stands high in the landscape because the resistant Massanutten sandstone is in the center of the fold. For much of its length, the mountain is actually two parallel ridges formed from the upturned edges of the syncline. The ridges tend to have fairly even summit elevations.

The northern third of this mountain contains a distinct synclinal valley running between the two ridges. You can enter this central valley from the north end between Front Royal and Strasburg. A gap visible from the highways for several miles on either side of Edinburg leads to the south end of the central valley. Although there is no gap in the ridge on the far side, a coach road once connected Edinburg and Luray.

The much larger New Market Gap is visible for many miles north and south of New Market. Here the Massanutten sandstone is breached and US 211 crosses the mountain on Martinsburg sandstone. Where the resistant Massanutten sandstone overlies the more easily eroded Martinsburg formation, undercutting of the sandstone permits talus to migrate downslope as barren areas visible along much of the mountain's flank.

The south nose of Massanutten Mountain is visible for many miles on either side of Harrisonburg. The Massanutten synclinal trough, however, continues southward to the vicinity of Greenville. Without the resistance provided by Massanutten sandstone, the fold hardly appears in the landscape.

To the west of the Shenandoah Valley is Little North Mountain, also with a comparatively even top. This ridge locates the structural front that divides the Shenandoah Valley to the east from the Allegheny Mountains to the west. The North Mountain fault, which thrusts older valley rocks upon younger Allegheny rocks, surfaces at Little North Mountain. South of Strasburg, you can see talus on the slopes of Little North Mountain similar to that on Massanutten Mountain. Beyond Little North Mountain to the west you can also see, on a clear day, Great North Mountain and other timbered ridges of the Alleghenies.

In the southern portion of the Shenandoah Valley, views of the Alleghenies are obstructed by knobs and ridges in the limestone rocks of the valley. Resistant zones of chert and of sandstone in the

Calcite columns, flowstone, and dripstone decorate the interiors of many of Virginia's 2500 caverns. Photo courtesy of Luray Caverns.

carbonate beds are responsible.

An exception to this generalization that sandstone and chert hold up the hills in the valley is Mole Hill about 5 miles west of Harrisonburg. With an elevation of almost 1900 feet, it rises some 500 feet above the surrounding valley floor. The resistant rock that supports this peak is a volcanic plug of olivine basalt. Since this volcanic rock formed during Eocene time, 50 million years ago, it is much younger than the host Ordovician Beekmantown formation, some 450 million years old.

Many roadcuts and outcrops along the Shenandoah Valley reveal bedding planes that dip steeply from their original horizontal attitude and trend very nearly parallel to the routes. Tectonic forces of the Alleghanian orogeny pushed these rocks into a series of open folds that are broken along various faults.

These limestones provide materials used in construction throughout this part of the Shenandoah Valley. Watch for them in house and barn foundations and in retaining walls. Although most of the retaining walls in Edinburg are made of locally quarried limestone, at least one along the main street is made of boulders of tan sandstone. Crushed limestone is road material beneath the streets and highways.

US 211
Warrenton — New Market
58 mi./93 km.

Warrenton stands astride the greenstone belt that marks the east flank of the Blue Ridge anticlinorium, a glorified anticline with lesser folds within it, that bounds the Blue Ridge geologic province on both sides in northern Virginia. This greenstone belt between highway 229 and Warrenton is the southeast flank of the Blue Ridge anticlinorium.

The topography cut into the Catoctin greenstone here is rolling, distinctly more rugged than that cut into the Pedlar massif to the west or the Triassic basin to the east. Few outcrops of greenstone appear along the road, but large blocks of vein quartz poke out of the grass in the fields on the north side of the highway between the Rappahannock River and Warrenton. Fresh excavations reveal the deep red color of the soil that forms on the greenstone.

Massive greenstone crops out in the median of US 211 west of the Rappahannock River. Rocks here tend to break into slabs along a cleavage defined by aligned needles of amphibole and flakes of mica. But you still see the original layering of the basalt flows in the bands of rock with their layers of gas bubbles. The southeast dip of the cleavage is much greater than that of the original flow layers.

The contact between the Lynchburg metamorphosed sediments and the overlying Catoctin greenstone is under the intersection with highway 229 south. Midway between Amissville and the Rappahannock River, the road cuts into the Lynchburg formation, here a pebbly

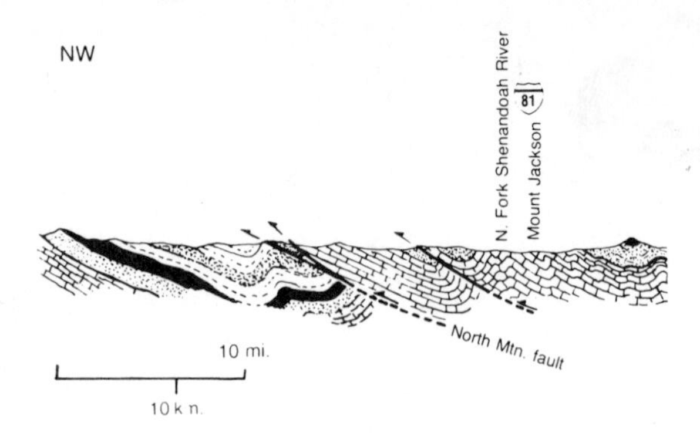

Cross section of Massanutten synclinorium in the Valley and Ridge geologic

sandstone with steeply dipping bedding. Some of the quartz grains, up to a pin head in size, are distinctly blue indicating a source area in the crystalline rocks now exposed just to the west, where similar distinctive quartz grains abound.

About one-half mile east of Amissville, dark gray sheared Robertson River granite full of big crystals crops out in the westbound roadcut. Gray gneiss crops out at the intersection with highway 621 in Culpeper County. The contact between crystalline basement complex and metamorphosed sedimentary rocks of the Lynchburg formation is about two miles east of Amissville.

Pavement outcrops in the fields and roadcuts along the route between Sperryville and Ben Venue provide abundant exposures of the Lovingston augen gneiss with its big crystals of pale feldspar. Business route US 211 through "little" Washington, where log cabins have been restored, has a roadcut in granite gneiss cut by a diabase dike at its western end. The wooded hill south of the highway about one mile east of Ben Venue is the north end of the Mechums River graben, a 55-mile-long by one- to two-mile-wide structure filled with sedimentary rocks thought to be the same age as those of the Lynchburg formation—late Precambrian.

The roadcut at the intersection of US 211 and US 522 in Sperryville is in Pedlar granite, here moderately sheared. Between the Piedmont Picnic Area, part way up the Blue Ridge, and Thornton Gap, the road crosses onto Catoctin greenstone. The contact is the Stanley fault. The tunnel at Thornton Gap is cut in Catoctin greenstone and the road continues in this formation down the west side of the Blue Ridge all the way to Shenandoah National Park Headquarters. The headquarters buildings are on the lowermost Chilhowee sandstone formations.

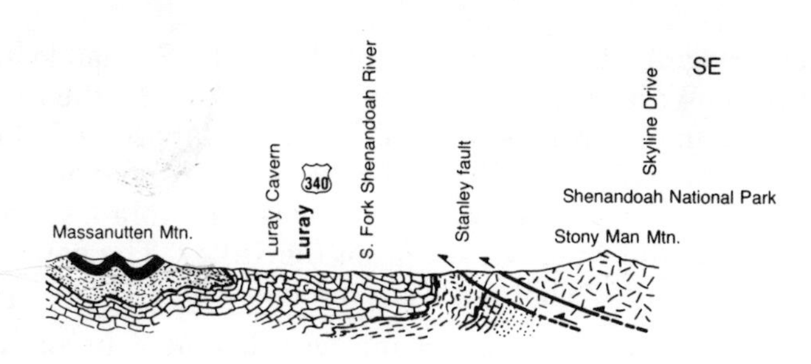

province from West Virginia through Luray to the Blue Ridge at Stony Man Mountain.

Thornton Gap was eroded through rock broken during movement of the Stanley fault. The facilities at Panorama sit squarely on this fault with Catoctin greenstone to the north and Pedlar granite to the south. From this overlook to the South Fork of the Shenandoah River, you see the Shenandoah Valley with Massanutten Mountain on the far side. The low pass in the mountain is New Market Gap, where US 211 crosses over to the North Fork of the Shenandoah River. To the left is the end of Massanutten Mountain, a syncline that continues as a fold but not as a mountain another 30 miles southwest. It is the core of resistant Massanutten sandstone that makes the fold stand up as a mountain.

From the park headquarters to Luray, noted for its caverns in the limestone, the route continues up the geologic section from the sandstones of the Chilhowee group into the rocks of the early Paleozoic carbonate bank. You are now in the Valley and Ridge geologic province.

Either bluff of the South Fork of the Shenandoah River affords excellent views of the Chilhowee foothills stretching into the distance to the southwest. From US 211 west of Luray and from the overpass over US 340, the southwestern nose of Massanutten Mountain is clearly visible when the air is clear.

Massanutten Mountain lies in the core of the wrinkled synclinorium of the Shenandoah Valleys. As you approach the mountain from either side, you travel up the geologic section to younger formations, the youngest being the highest in elevation. Conversely, as you come down off the mountain on either side, you travel down the geologic section.

Views northwest from New Market Gap include the valley of the North Fork of the Shenandoah River with Little North Mountain, the sharp ridge, on its northwest flank. Beyond is Shenandoah Mountain with the West Virginia state line on its crest. Southeast is the Blue Ridge with its sharp foothills of Chilhowee sandstone forming the far side of the valley of the South Fork of the Shenandoah River. The barren, blocky peak across the valley is Stony Man, so named for its human profile when viewed from the valley. The gap to the north beyond Luray is Thornton Gap through which US 211 passes.

Although the carbonate bank was deposited in shallow, sunny water, the Martinsburg formation was deposited in deep, dark water. The rock you see in New Market Gap is thick beds of sandstone not typical of the formation as a whole. Each bed was a single surge of sand, some more than two feet thick, into the suddenly deepened sedimentary basin. Note that the beds dip in toward the mountain on both sides, thus revealing the synclinal structure of the ridge.

New Market Gap exists because the resistant Massanutten sandstone is here breached with the gap cut into the underlying Martinsburg formation. Structurally, this is a high point in the Massanutten synclinorium.

New Market is in the valley of the North Fork of the Shenandoah River, between Little North Mountain on the west and Massanutten Mountain on the east. The valley here is eroded into rocks of the early Paleozoic carbonate bank of Cambrian and Ordovician age. Knobs and hills in the valley are underlain by cherty or sandy portions of the carbonate bedrock.

Thick beds of red sandstone of the Martinsburg formation crop out in the roadcut at New Market Gap. Horizontal bedding here marks the axis of the Massanutten syncline.

The Chopawamsic formation created rapids along the Fall Line at the upper end of Abel Reservoir.

IV
THE FALL LINE

The Fall Line, sometimes called the Fall Zone, is an important geologic and economic boundary. Its name comes from the bedrock rapids where stream channels cross it. In Colonial Times, before siltation of channels resulting from farming and soil erosion, the Fall Line was the upstream western limit for ocean-going ships. It was also the downstream eastern limit for crossing rivers without need to resort to ferrying cargo. Trade dictated that commercial centers be located on the rivers of Virginia at the Fall Line.

Estuaries extending from the Fall Line to Chesapeake Bay are fresh at their upstream end and salty toward the bay. Since barnacles and other marine growth on wooden hulls cannot tolerate fresh water, a visit to one of the Fall Line ports offered the additional incentive to ship captains of a free bottom cleaning.

Geologically, the Fall Line lies at the western edge of Coastal Plain sediments lapped onto the crystalline igneous and metamorphic rocks of the Piedmont geologic province. But rock formations of the Piedmont extend seaward beneath the Coastal Plain sediments. Indeed, some quarries in the Petersburg granite are within the Coastal Plain geologic province; the thin veneer was scraped off to expose the granite

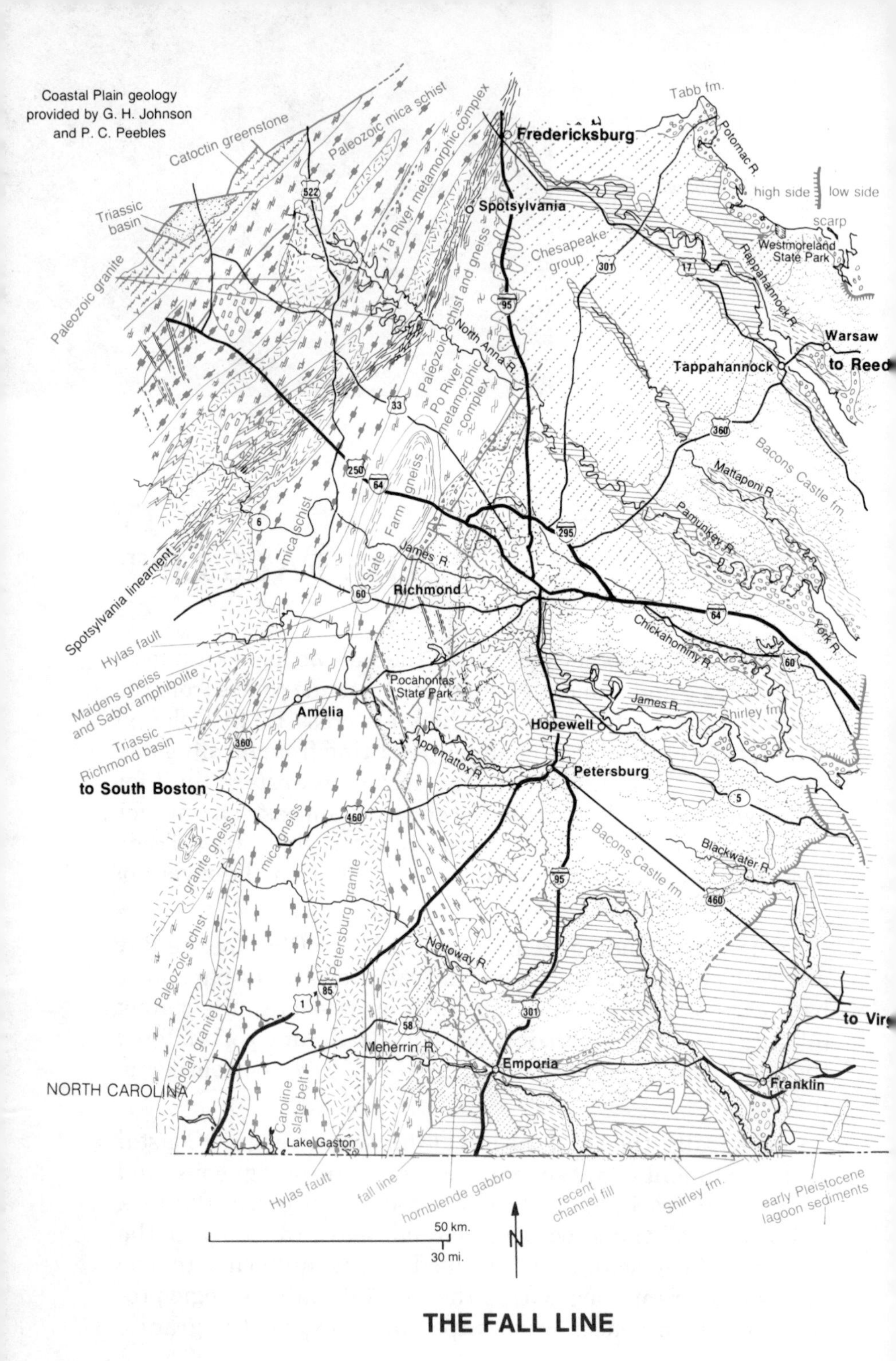

THE FALL LINE

beneath. Drilling and seismic exploration reveal that Piedmont rock formations extend under the entire Coastal Plain province.

The Coastal Plain onlap once may have been tens of miles west of its present location. Geologists interpret some of the eastern Piedmont topography as exhumed from a cover of sediments because some of the stream patterns otherwise make no sense. Many present-day streams flow near but not in bedrock channels.

JAMES RIVER AT THE FALL LINE

The city of Richmond straddles the James River where it crosses the Fall Line. Here, the Coastal Plain of Cretaceous to Recent sedimentary formations laps onto the crystalline rocks of the Piedmont. In Colonial times, this was the head of navigable waters; downstream the James River is an estuary of Chesapeake Bay.

Rocks of the Piedmont province are deeply weathered to saprolitic clay. You see bedrock only in stream channels and quarries. The saprolite is itself covered in much of the eastern Piedmont by terrace deposits laid down by streams when their channels were at higher elevations.

Bedrock under Richmond is the Petersburg granite which invaded Precambrian age Piedmont rocks about 330 million years ago, during Mississippian time. Most of the stone quar-

The Richmond skyline rises above the Lee Memorial Bridge over the James River.

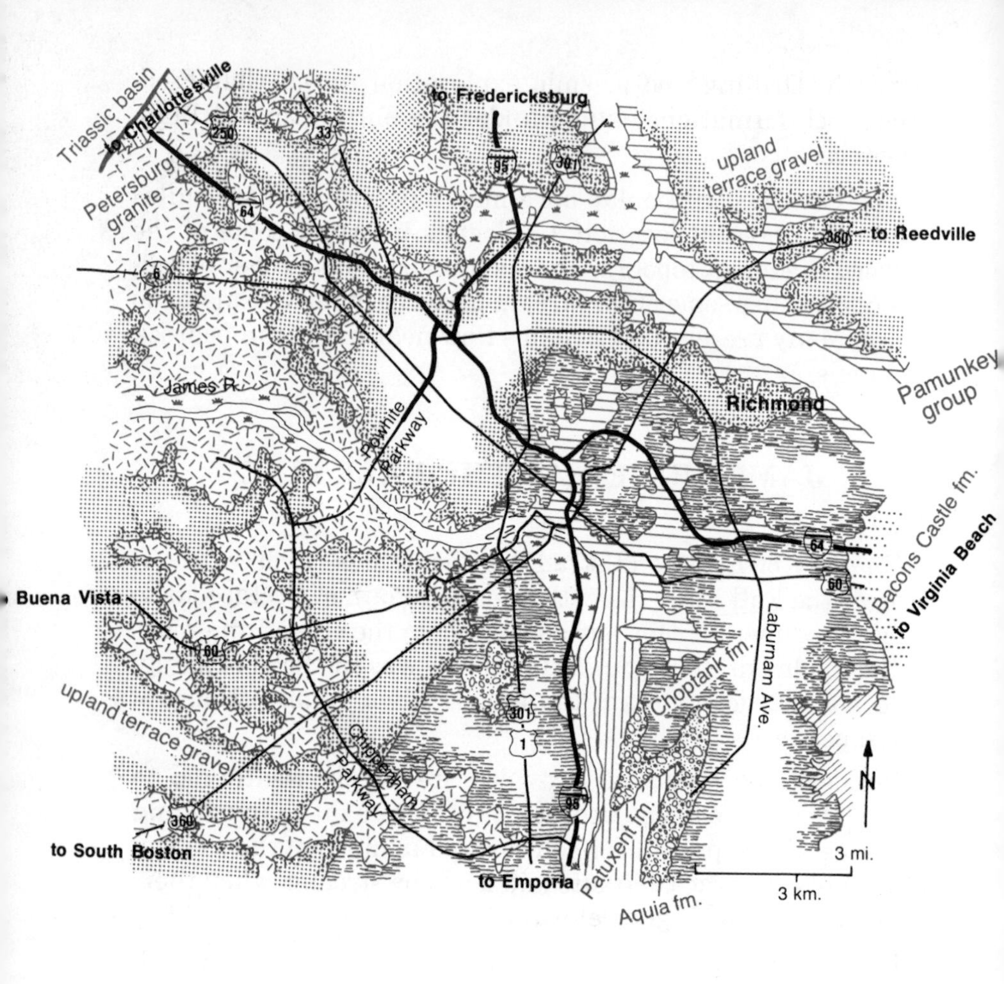

JAMES RIVER AT THE FALL LINE

ries in the Richmond area worked this pink granite, formerly much used for dimension stone. One of the best places to see some of the varieties of the Petersburg granite is downstream from James River Park off US 1 at the Robert E. Lee Bridge. The granite here is in places massive, in others streaky because it has a foliation. Zones of coarse-grained pegmatite are also present. Pictures of the marvelous stream potholes here have decorated geology textbooks for over a century.

View upstream along the James River at the Fall Line.

Prized as a dimension stone for buildings and monuments, the Petersburg granite held at one time more than 18 operating quarries. The Petersburg granite is still quarried in the Richmond area, primarily for crushed stone for road metal and concrete aggregate.

When rifting of Pangaea began to form the Atlantic Ocean during Triassic time, about 220 million years ago, great fractures formed in the Paleozoic rocks of the present Piedmont geologic province. Between these fractures the rock slowly dropped down to form basins that filled with water and sediment. In Jurassic time, basalt dikes intruded these basins. The Richmond and Taylorville basins form the western boundary of outcrops of the Petersburg granite.

Thinning of the Earth's crust adjacent to the widening Atlantic Ocean permitted deposition of marine sediments on the eroding surface of the Petersburg granite. The oldest post-Atlantic material in the Richmond area, however, includes stream and possibly estuarine gravels, sands, and silts, which cap the uplands west of downtown Richmond. These sediments protected the underlying material from erosion and so caused the topography to become inverted—the former stream channels are now the higher areas between modern channels.

In Miocene and Pliocene times, Richmond was flooded by the Atlantic Ocean, which then deposited sand, silt, and clay of the Chesapeake group of formations. These are the surface deposits from Mechanicsville to Lakeside. Ocean water receded from the area in middle Pliocene time, perhaps to more than

100 feet below present sea level. Streams then flowing off the Piedmont dug their channels to bedrock in the Petersburg granite.

Late Pliocene seas returned and deposited gravel, sand, and silty clay of the Bacons Castle formation. This sea level was not as high as its predecessor, however, and the Bacons Castle formation lies at a lower elevation than the Chesapeake group on the erosional surface carved in middle Pliocene time. Bacons Castle sediments are the surface deposit on the uplands in downtown Richmond as well as to the east and south.

When sea level fell again, the James River established its present valley, if it had not done so earlier. During Pleistocene time, each glacial advance to the north caused sea level to fall and the James River to scour its channel. Each interglacial period brought an estuary into the James River valley, along with deposition of gravel, sand, and silt.

During each of these marine advances, the ancient shoreline was somewhere in the general vicinity of the present Fall Line, perhaps even ten or twenty miles west of it. Geologists interpret stream channel patterns on the eastern Piedmont as evidence that the rock was exhumed by erosion from beneath Coastal Plain sediments. At times, this area may have looked much like the modern coasts of Connecticut and Rhode Island.

During each low stand of sea level, the area around Richmond must have looked much like it does today. Rivers draining off the Piedmont surely cut valleys in a gently rolling to nearly flat topography. Only the vegetation and animals would have been different.

For most of the time, however, these rivers would not have swamps bordering their channels, as we see today. Our modern swamps result from partial advance of the sea onto the continental margin. Estuaries fill the seaward ends of the large rivers while swamp deposits have partly filled the small rivers. Should the world's remaining ice caps and glaciers melt, the sea would again advance to the neighborhood of the Fall Line, beginning another round of sediment deposition. Should another ice age get underway, sea level would fall and the eroding rivers would quickly drain their bordering swamps.

US 1, US 301, Interstate 95
Fredericksburg — North Carolina state line
133 mi./214 km.

The route follows the Fall Line for its entire distance. Uplands are underlain by Coastal Plain sediments deposited during Cretaceous and Tertiary times, while the major streams and rivers have eroded their beds into the underlying igneous and metamorphic rocks of the Piedmont geologic province. Downtown Fredericksburg is built on sands, clays, and gravels of the Cretaceous Patuxent formation. The interstate highway at Fredericksburg skirts the western edge of the sedimentary onlap except where rivers have cut through to crystalline rocks of the Fredericksburg complex.

The Rappahannock River is one that cuts through to the Fredericksburg complex, here biotite gneiss of the Po River metamorphic suite, visible from US 1. The Fall Hill fault crosses under the interchange between the Interstate highway and Virginia 3. This fault divides the older Po River metamorphic suite from the younger Ta River metamorphic suite, chiefly dark amphibolite. Both are about 500-600 million years old.

Between the North Anna River and the South Anna River, the routes cross onto the Taylorsville Triassic basin about where it passes beneath the sediments of the Coastal Plain. Between the Taylorsville basin and the James River, the route is on Miocene marine sediments which lie on the 300-million-year-old Petersburg granite. The James River has cut through the Tertiary sediments, that now form bluffs on

either side, and into Petersburg granite.

Between Richmond and Petersburg, the routes are on Tertiary sediments, but the Appomattox River has cut through to Petersburg granite. South of Petersburg, the sediments are the Pliocene Bacons Castle formation. The Nottoway River has cut into the Bacons Castle formation but is now choked with Pleistocene swamp deposits. The Meherrin River has cut through the Tertiary sediments and into dark hornblende gabbro. At the North Carolina state line the routes cross more Petersburg granite.

US 1, Interstate 85
Petersburg — North Carolina state line
69 mi./111 km.

For several miles southwest of Petersburg, the route runs on Tertiary sediments except where streams have cut through to the underlying Petersburg granite, which intruded the Piedmont about 330 million years ago. The thinness of this sedimentary cover may be judged from the presence of a granite quarry just west of the US 460 exit. Little sediment needed to be scraped away to get to the igneous rock.

Near Dinwiddie, the Tertiary sediments have been eroded off and the route crosses weathered Petersburg granite. Between there and the Nottoway River, the route crosses a band of phyllite and an intrusion of hornblende gabbro before returning to the Petersburg granite. Between the Nottoway and Meherrin rivers there are narrow bands of mica gneiss, volcanic rock, another band of Petersburg granite, and more mica gneiss. At Lake Gaston the route crosses an outlier of Redoak granite of uncertain age.

Most of these formations are so deeply weathered to saprolite that there is little evidence of their character beyond soil color in excavations and freshly plowed fields. The deeper the red of the fresh exposure, the more iron was in the original rock. Exceptions to deep weathering are the natural rock gardens of Petersburg granite along US 1 between Dinwiddie and DeWitt.

Although most granite tends to weather to topographic low areas under the Piedmont province, some unfractured Petersburg granite forms distinctive knobs as in this lawn along US 1 southwest of Petersburg.

US 301
Potomac River — Richmond
69 mi./111 km.

The route enters the state on the Virginia side of the Potomac River, which lies entirely in Maryland. Here the flood plain is cut into the late Pleistocene Tabb formation. Where the route rises from creek and river bottoms, it passes onto older Tertiary marine sediments.

For a short distance south of Bowling Green, the route parallels the Mattaponi River, formed below the confluence of the Mat, the Ta, the Po, and the Ni rivers. Where it crosses this river, the channel is choked with Pleistocene sediments of the Tabb formation, as are the Pamunkey and Chickahominy channels.

Between Hanover and Richmond the route is on Tertiary marine sediments older than the Bacons Castle formation. Petersburg granite can be seen in the channel from the Lee Memorial Bridge, where the James River cut through the Tertiary and Cretaceous formations of the Coastal Plain. The bluffs on both sides of the river are capped by Tertiary sediments.

Two potholes in the Petersburg granite at the Fall Line in Richmond.

US 360
Reedville — Richmond
86 mi./138 km.

The route begins in Reedville on the Wicomico River near its mouth on Chesapeake Bay. Sediments and beach deposits here are the Tabb formation. The bluffs to the west are underlain by sediments of the older late Pliocene Bacons Castle formation. Just west of Reedville, still older Cretaceous sands and silts appear at the surface. Most of the upland is covered by Tertiary terrace sands and gravels.

Cretaceous sandstone appears in roadcuts on the west bluff of the Rappahannock River. West of Tappahannock, the road travels on Tertiary marine sediments older than the Bacons Castle formation. Channels of the Mattaponi, Pamunkey, and Chickahominy rivers, all much deeper during Pleistocene ice ages, are backfilled with late Pleistocene and recent sediments.

From the bridge in Richmond, you can look down at the Petersburg granite in the channel. The James River eroded through the overlying Tertiary sediments exposed on either bluff into the 330-million-year-old granite.

US 17
Fredericksburg — Tappahannock
47 mi./76 km.

Like Petersburg and Richmond, Fredericksburg sits athwart the Fall Line, named for the many rapids and riffles along this north-south zone. The Fall Line marked the head of navigation for ocean-going vessels and the easternmost location for north-south land travel without ferry crossings in Colonial times. Such intersections were destined to become centers of trade. The Fall Line also marks the downstream extent of bedrock channels and the beginning of tidal estuaries in the Coastal Plain.

From Fredericksburg to Tappahannock, the route follows the Rappahannock Estuary. Upland surfaces here are underlain by Tertiary marine sediments, part of the Coastal Plain. The valley, which was much deeper when sea level was lower, is now filled with Pleistocene to recent fluvial and estuarine sediments.

US 58
Virginia Beach — Emporia
98 mi./158 km.

Most of the route from Virginia Beach through Norfolk is on sediments of the late Pleistocene Tabb formation. Sediment at the surface of Dismal Swamp is also recent. Suffolk is on the west side of the Dismal Swamp, at its north end.

The sediments under the west or up side of the Suffolk scarp belong to the middle Pleistocene Shirley formation while those under the east or down side are the considerably younger Tabb formation. The scarp itself is an old shoreline that existed while the Tabb formation was accumulating.

Between Suffolk and Courtland, the route crosses a plain underlain by unconsolidated lagoonal sediments laid down during early Pleistocene time. Sea level then ranged from 50 to 100 feet above its present stand. The Blackwater, Nottoway, and Meherrin rivers cut their valleys through these sediments during times of low sea level and are now filled with recent channel and swamp deposits. Between Courtland and Emporia, the uplands are underlain by older Pliocene sediments of the Bacons Castle formation.

Emporia is on the Meherrin River at the Fall Line—the contact between the crystalline rocks of the Piedmont geologic province and sediments of the Coastal Plain geologic province. Beginning some 10 miles west of Emporia, the uplands on either side of the river are underlain by Tertiary sediments capped by a veneer of upland sands and gravels. The Meherrin River has cut its valley through these sediments to an underlying hornblende gabbro as far east as Emporia.

US 360
Richmond — South Boston
113 mi./182 km.

Richmond is on the James River where Coastal Plain sediments lap onto the much older crystalline rocks of the Piedmont. Southwest of Richmond, the route crosses Tertiary sediments laid down on Petersburg granite.

This small granite quarry west of Amelia is cut by many small pegmatite dikes.

The pale pile beyond the Morefield pegmatite mine head contains recently-dug green amazonite, a variety of feldspar.

About 12 miles southwest of the James River, the route crosses onto the Richmond basin with its fill of Triassic and Jurassic sedimentary rocks. Swift Creek Reservoir is within the basin. The route crosses the western border fault of the basin at the bridge over the Appomattox River. Between the river and Jetersville the bedrock is mica schist and mica gneiss, Piedmont rocks.

Granite pegmatites of the Amelia district have been mined for mica. They have also yielded some remarkable mineral specimens, including moonstone and amazonite varieties of microcline feldspar. Although most of these pegmatites are no longer worked commercially, several of the present owners permit collectors to rummage, for a fee, for minerals in the mine dumps and abandoned quarries.

Pegmatite dikes, some more than 50 feet thick and 1,300 feet long, cut across the foliation of the biotite gneiss host rock. White muscovite mica in booklets more than a foot across provided the primary objective in development of these pegmatites in the last century. When it was used in stove and furnace windows, all pieces smaller than two inches square were discarded. Later, with the advent of electrical appliances requiring insulation, all the mica became valuable.

Quartz from the pegmatites was shipped for use as an abrasive. Most of the feldspar went into the ceramic industry, although nicely colored green amazonite variety of feldspar was prized as a semiprecious gem. Feldspar had altered to kaolinite clay above the water table and that was used to make china dinnerware.

In addition to fees from mineral collectors, modern commercial interest in these pegmatite bodies exploited the masses of amazonite, the green semiprecious gem variety of the feldspar mineral microcline. During the 1960s, amazonite from the Rutherford Mine was exported to Germany for carving. In 1985, the Moorefield Mine was re-opened with the same objective.

Pegmatite was also quarried 8 miles southwest of Amelia at Jetersville where the host is granite. This granite is also cut by Triassic basalt dikes and older hornblende gabbro dikes in the Burkeville area.

Between Burkeville and Green Bay, the route passes from the granite onto mica schist which underlies it to Briery. From just south of Briery almost to Wylliesburg, the road crosses volcanic rock. Wylliesburg is on the Redoak granite, which intrudes the Virgilina synclinorium to the south. Although best known for its past copper mining, the Virgilina district also was the scene of gold mining. The road crosses a corner of this district at the bridge over the Roanoke River.

From the Virgilina district to South Boston, US 360 and highway 340 cross granite gneiss just west of a small Triassic basin. Here the road roughly follows the Spotsylvania lineament, an enigmatic line across the state that divides areas in which the magnetic maps have distinctively different patterns.

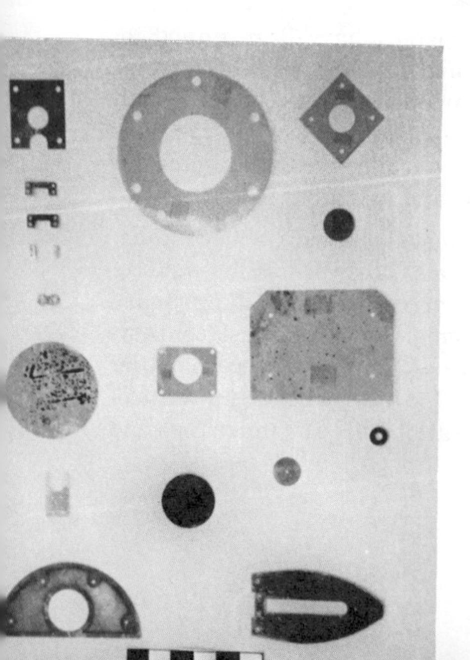

Large mica booklets from pegmatites around Amelia and in North Carolina were split into thin sheets, then cut into shape for a variety of commercial applications where thermal or electrical insulation was required.

V
BLUE RIDGE AND SHENANDOAH VALLEY

THE CENTRAL APPALACHIAN MOUNTAINS

The Appalachian chain of ridges and valleys extends from Alabama to Nova Scotia. Roanoke lies near the division between the central and southern sections of this range. The northern end of the central section is near New York City.

The earliest major geologic event recorded in the Appalachians between Roanoke and the Potomac River is the Grenville metamorphism and igneous activity in the Blue Ridge. That happened about 1200-1000 million years ago. Although this event obliterated most of the rock record of earlier happenings, older rocks had to be present to be metamorphosed. These earlier rocks, now granulite gneisses, may have been interbedded volcanic and sedimentary rocks with a history going back beyond 1800 million years.

Igneous rocks of Grenville age are charnockites, unusual feldspar-rich rocks with variable amounts of blue quartz. They differ from common granites in that they contain pyroxene and garnet. These two groups of minerals suggest that the rocks crystallized in a crustal environment that had less water

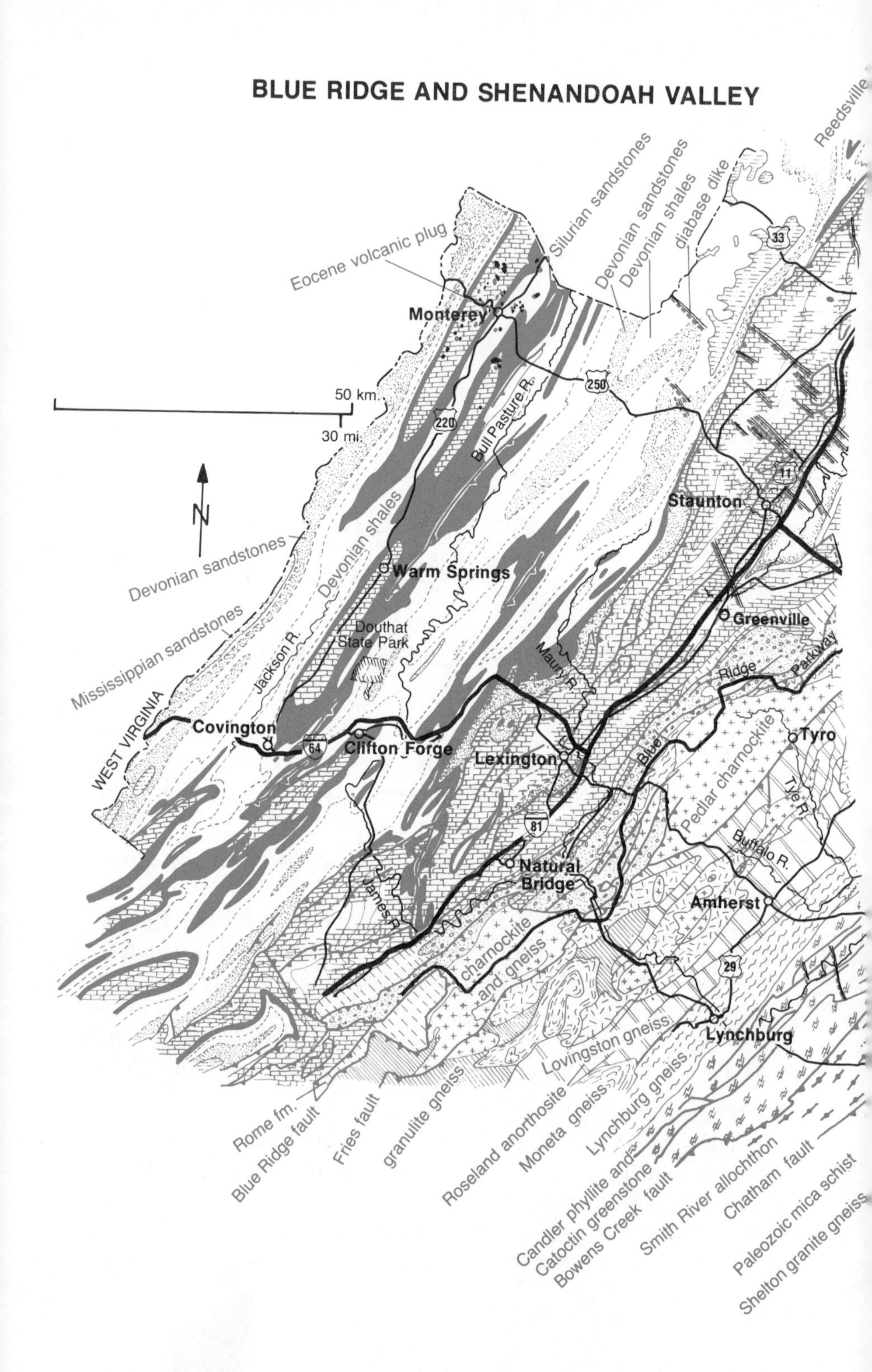
BLUE RIDGE AND SHENANDOAH VALLEY
Reedsville
Silurian sandstones
Devonian sandstones
Devonian shales
diabase dike
Eocene volcanic plug
Monterey
33
250
50 km.
30 mi.
220
Bull Pasture R.
11
Staunton
N
Devonian shales
Devonian sandstones
Warm Springs
Mississippian sandstones
Jackson R.
Douthat State Park
Greenville
Maury R.
Ridge
Parkway
WEST VIRGINIA
Covington
64
Clifton Forge
Lexington
Blue
Tyro
Pedlar charnockite
Tye R.
81
Buffalo R.
Natural Bridge
Amherst
James R.
charnockite and gneiss
29
Lynchburg
Lovingston gneiss
Rome fm.
Blue Ridge fault
Fries fault
granulite gneiss
Roseland anorthosite
Moneta gneiss
Lynchburg gneiss
Candler phyllite and
Catoctin greenstone
Bowens Creek fault
Smith River allochthon
Chatham fault
Paleozoic mica schist
Shelton granite gneiss

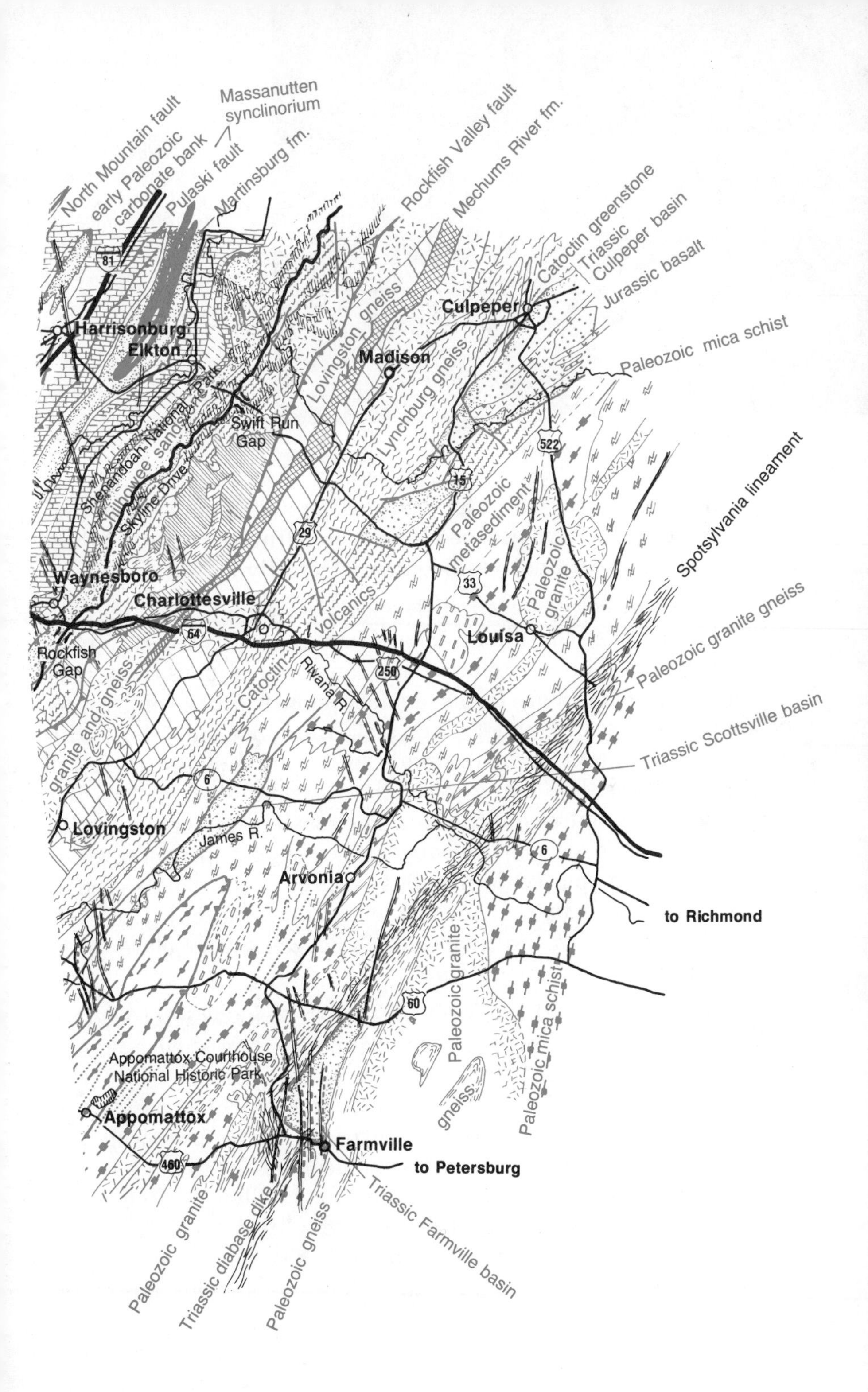
Massanutten synclinorium
North Mountain fault
early Paleozoic carbonate bank
Pulaski fault
Martinsburg fm.
Rockfish Valley fault
Mechums River fm.
Catoctin greenstone
Triassic Culpeper basin
Jurassic basalt
Paleozoic mica schist
Spotsylvania lineament
Paleozoic granite gneiss
Triassic Scottsville basin
Harrisonburg
Elkton
Culpeper
Madison
Lovingston gneiss
Lynchburg gneiss
Shenandoah National Park
Chilhowee sandstone
Skyline Drive
Swift Run Gap
Paleozoic metasediment
Paleozoic granite
Louisa
Waynesboro
Charlottesville
Catoctin volcanics
Rockfish Gap
granite and gneiss
Rivana R.
Lovingston
James R.
Arvonia
to Richmond
Paleozoic granite gneiss
Paleozoic mica schist
Appomattox Courthouse National Historic Park
Appomattox
Farmville
to Petersburg
Paleozoic granite
Triassic diabase dike
Paleozoic gneiss
Triassic Farmville basin
81
64
29
15
522
33
250
6
60
460

Nellysford granulite gneiss here shows very tight folds. These folds actually refold older folds.

available than is typical for granites. Or any dissolved magmatic water may have escaped before they finished crystallizing. The problem has not yet been satisfactorily solved.

The charnockites around Roseland between Charlottesville and Lynchburg contain an intrusion of the relatively rare rock anorthosite. This enigmatic rock, composed of more than 90% plagioclase feldspar, forms an intrusion 9 miles long by as much as 2 miles wide. This rock was once mined for the titanium content of the minerals rutile and ilmenite, which it also contains. For many years the Roseland district was the world's chief source of titanium. The rock is a geological puzzle in that its composition indicates higher temperature of crystallization than you would normally expect in the Earth's crust. And its crystals up to three feet across indicate that it crystallized slowly, with the temperature hovering very close to its melting point for a long time.

For the next 500 million years these rocks were eroded by wind and running water. The sediments they shed accumulated on an old continental margin somewhere not far east of the present Blue Ridge. We call them the Lynchburg and Candler formations. Late in this history, rifting produced the Batesville and perhaps other fault-block basins much like the younger Triassic basins farther east. These filled with shallow-water sediments, the Mechums River formation. The Swift Run formation may be another basin fill, or perhaps it is sediment caught in transit in valleys in ancient Grenville mountains.

Another pulse of igneous activity took place about 600 million years ago, in Eocambrian time. Overheating in the Earth's mantle caused melting and swelling below and rifting above. Magma rose to the base of the crust but halted there because it was too dense to rise through the overlying granitic crust. There it dissolved low-density minerals while crystallizing out high-density minerals, eventually becoming basalt. This basalt then rose through the crust in fissures and spilled out on the eroded surface of Grenville age rocks and the sediments of the Swift Run, Lynchburg, and Candler rocks. We call these basalt lava flows and the layers of sediment between them the Catoctin formation. The basalt left in fractures crystallized as coarser diabase dikes that form minor ridges in the Blue Ridge complex.

While mantle magma was adjusting its density for its trip through the continental crust, it was losing heat to the overlying crustal rocks, which melted to form granitic magma. Some of this crustal magma erupted to form rhyolite in the northern end of the Blue Ridge; most intruded overlying charnockites and granulites before crystallizing into granite. The largest of these intrusions is the Robertson River granite, which has an outcrop area 70 miles long by less than 5 miles wide.

Diabase dike, now amphibolite, contains blebs of granite that indicate two magma sources.

Charnockitic granite typically crops out as rounded knobs.

West of the Blue Ridge, vulcanism gave way to a thick sequence of sandstones now known as the Chilhowee group. Evidence from ripple marks and cross beds indicate that much of the sand in those rocks was in beaches and shoals near the North American shore of the Iapetus Ocean, the ancestral Atlantic. As the western source area for the Chilhowee group eroded, the sands became purer and generally finer grained than the earlier deposits, which were pebbly and full of feldspar. Sparse fossils indicate that the upper portions of the Chilhowee sandstones, at least, were deposited in early Cambrian time.

From early Cambrian to late Ordovician time, a span of 100 million years, eastern North America was a shallow tropical sea. That was just the right environment for deposition of the early Paleozoic carbonate bank. A few beaches survive as sandstones in the sequence of limestone and dolomite. During a few million years in middle Ordovician time, much of the carbonate bank was above sea level and dissolving in the rain. This uplift may have been associated in some way with the beginning of closure of the Iapetus Ocean.

Then something caused the crust under eastern North America to bend down. Some geologists think this something was the arrival of the Inner Piedmont belt of the Carolinas and the out-of-place Smith River allochthon of the Virginia Piedmont. Those scraps of continental crust docked onto North America as the Iapetus Ocean slowly closed. Whatever the reason, there is no doubt that a flood of sediments poured onto the depressed edge of the continent from some source to the east, presumably mountains raised by the Taconic orogeny. By early Silurian time, the trough had filled and the area was covered by a great sand beach and shoal deposits that became the Tuscarora and Massanutten sandstone formations.

The next 40 million years, from late Silurian to early Devonian time, were quiet. A middle Paleozoic carbonate bank slowly grew on the eastern shelf of North America. But the Iapetus Ocean was still closing, bringing island arcs and microcontinents toward the American shore. In middle Devonian time another terrane, perhaps now hidden by Coastal Plain and continental shelf sediments, docked on the North American continent. The continental margin sank again. Clays destined to become the Devonian shale formations flooded into the newly formed trough from mountains built during the Acadian orogeny.

In Mississippian and Pennsylvanian time, tropical swamps farther west were periodically drowned and covered with clay. Peat deposited in the swamps later became coal seams. If these formations were also deposited in the central Appalachians of Virginia, they have long since eroded away.

Except for possible stray pieces of oceanic crust caught up in the Piedmont province or its extension under the Coastal Plain, the Iapetus Ocean ceased to exist during the Alleghanian orogeny of about 320 million years ago, toward the end of Pennsylvanian time. What is now Africa and Europe crunched onto North America then with a great shove that buckled the rocks into the folds and faults we see today in the Blue Ridge and Valley and Ridge provinces.

The Blue Ridge anticlinorium, the Massanutten synclinorium, and all the lesser anticlines and synclines of the Valley and Ridge province, formed as the rocks buckled under pressure from the southeast. The Blue Ridge was shoved over rocks of the Shenandoah Valley on a thrust fault. And the rocks of the Shenandoah Valley shoved over an entire stack of rocks of the Valley and Ridge along another thrust fault. Rock formations from Cambrian to Silurian in age are stacked twice under the Shenandoah Valley.

The Alleghanian orogeny also caused minor metamorphism of the Blue Ridge rocks. Sandstones escaped almost unscathed, but shales and some volcanic rocks became phyllites. Veins of quartz, calcite, and pyrite filled fractures in the rocks.

In Jurassic and again in Eocene times, volcanic activity in the form of isolated volcanoes, dikes, and sills emerged in the west-central Valley and Ridge province. Mole Hill west of Harrisonburg and Trimble Knob in Monterey are among the results. The thin basalt sill near the base of the pinnacles of limestone in Natural Chimneys Regional Park is another.

GORGES AND GAPS

For most of their courses in the Valley and Ridge geologic province, rivers flow between parallel ridges of resistant sandstones. The valley may be in the core of a breached anticline or syncline, or it may be cut into an easily erodible formation on the flank of a fold. Hightown Valley is a breached anticline; the headwaters of the Jackson River are in a syncline; and both forks of the Shenandoah River are in flank valleys on either side of the Massanutten synclinorium.

The Maury River cuts through ridges of Tuscarora sandstone at Goshen Pass.

Gorges formed where rivers cut across the trend of the rocks and breached one of the resistant formations, most commonly sandstone. Flow in these water gaps is turbulent over bedrock and boulders. But why do streams flow through "tough" ridges that, in many cases, they might have flowed around following much "easier" courses?

The answer is that these valleys were originally cut at higher levels in a rock, now removed by erosion, that was different in composition and structure from that now exposed. Rivers have no way of knowing the geologic complexities that lie in wait below them. As erosion reduces the general level of the landscape, streams commonly come down onto hard rocks that were not exposed when they started to flow. Pennington Gap through the overturned northwest limb of the Powell Valley anticline and the James River gorge through the Blue Ridge formed where rivers eroded down onto resistant rock formations.

Relatively smaller streams with less erosive power didn't cut down as fast as their larger neighbors. Their flow was eventually captured by larger streams eroding weakly resistant rocks. Remnants of the old valleys of smaller streams survive as notches in ridge crests. Cumberland Gap is just such a wind gap left when the Powell River decapitated the Cumberland River and pirated her upper reaches. Many of the mountain passes in the Valley and Ridge and in the Blue Ridge geologic provinces are similar segments of abandoned valleys.

YOU CANNOT DRAW A CHECK ON A CARBONATE BANK

Limestone and dolomite are the two most common sedimentary carbonate rocks. Metamorphosed, both become marble, which we have in Virginia. Limestone is composed primarily of the mineral calcite, calcium carbonate; dolomite is primarily the mineral dolomite, calcium magnesium double carbonate. Geologists can tell crystals of one from the other, but in the field, in a cliff or a roadcut, rocks composed of these minerals may be all but impossible to tell apart.

Small stream cascades over layers of flat-lying limestone.

Since the distinction between limestone and dolomite is difficult, the sedimentary carbonate rocks that appear along the highways are lumped together in this book as limestone or carbonate. But a geologist looking for limestone to make Portland cement must know the difference because dolomite's magnesium will ruin the concrete.

Most limestone along the highways in the Valley and Ridge province was deposited on a bank of a shallow tropical sea such as now surrounds the Bahama Islands, the modern Bahama Bank. Three separate carbonate banks are preserved in the rock record in Virginia: of early, middle, and late Paleozoic age. All three contain both limestone and dolomite and certain individual formations change from limestone in the east to dolomite in the west.

From early Cambrian through middle Ordovician times, a period of 100 million years, the early Paleozoic carbonate bank stretched from Pennsylvania to Texas and from Virginia to Illinois. Geologists divide this carbonate bank into as many as seven different formations based on differences in the rocks and fossils. But many look so much alike that formation names are little used in this book.

A second carbonate bank, the middle Paleozoic one, was less extensive and lasted only from upper Silurian to lower Devonian times, about 30 million years. This bank contains as many as four formations, also not readily distinguished except by the trained eye.

A third carbonate bank, the late Paleozoic one, was even less extensive and lengthy and less permanent. Confined to the westernmost regions of Virginia, this bank was interspersed with sandy sediments and coal measures. It lasted about 20 million years until the Alleghanian orogeny and the closure of the Iapetus Ocean drained the eastern portion of the continent.

The carbonate banks, like the modern Bahama Bank, accumulated in warm shallow seas teeming with the life of the times. They record long periods of crustal stability with relatively small, tens of feet, change in water depth. The dearth of clay and sand formations indicates that the land above sea level was very low and shedding little weathered rock into the streams flowing into this sea. Magnetic and fossil evidence indicate the existence of deep ocean to the east.

The carbonate banks ended very suddenly, geologically speaking. The first two went out of business as deep ocean replaced the shallow sea. Mountains were forming elsewhere in the Appalachian chain at these times, so the sudden deepenings are part of that picture: first the Taconic, second the Acadian orogeny. The third carbonate bank ended with closing of the Iapetus Ocean, when mountains rose where before there had been seas. Even though they all eventually went out of business, these carbonate banks lasted longer and were more stable than any that deal with money.

HOW TO MAKE A CAVERN

A cavern is just a hole where carbonate rock—limestone, dolomite, or marble—has been disdolved away by acid ground water percolating through the formation. Where does this acid come from? Although carbonic acids from atmospheric carbon dioxide and humic acids from decaying vegetation contribute to groundwater acidity, the main villain is sulfuric acid derived from oxidation of the sulfide mineral pyrite. Many of the limestone formations contain small amounts of pyrite and shale may contain a lot of it.

Solution of limestone leaves small to large cavities, which are then potential sites for precipitation from solution of carbonate minerals. How large do these cavities get? Before it collapsed and filled to its present form, the cavern that became Natural Chimneys could have housed a 40-story office building with room to spare.

Carbonate rock presents special problems if you live on it or get your drinking water from it. People living in areas under-

Ground water etches limestone bedding and solution cavities permit collapse breccia.

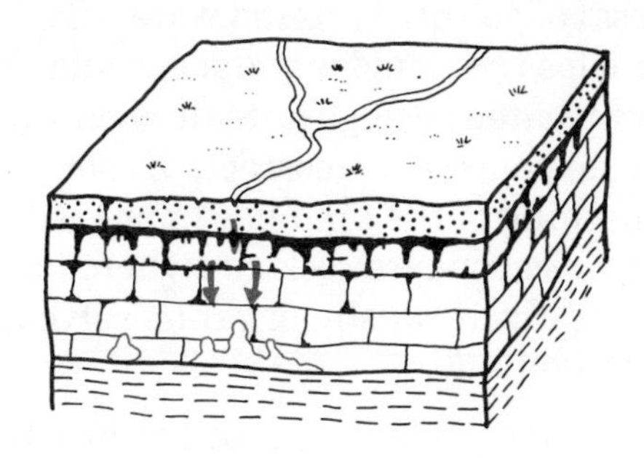

Acid water percolates down through limestone and dissolves the rock along fracture joints. Insoluble clays in the limestone and ceiling blocks loosened by solution accumulate on the floor of a cavern. Eventually unsupported roof caves in to give you a sinkhole in the ground. Many sinkholes lead into cavern systems.

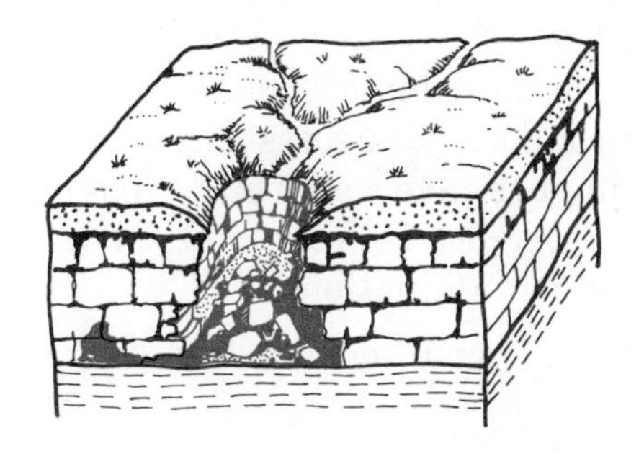

lain by carbonate rock can benefit or suffer from what can happen to that bedrock. One advantage is the huge volumes of water that can be pumped from a well that taps a developed, dissolved-out fracture joint system below the water table. But dissolved chemicals from carbonate rock make the water hard. This is a disadvantage for washing and can inspire you to buy a water softener. But there are some indications that hard water may somehow help prevent certain heart diseases.

Two major hazards of areas underlain by carbonate rock are cavern collapse, producing new sink holes, and lack of filtration permitting wide dispersal of polluted ground water. Buildings dropped into new sinkholes that suddenly appear and the unexpected appearance of untreated sewage or toxic chemicals in once potable water are expensive to correct, if not immediately fatal. Although nothing of the kind has happened so far in Virginia, houses built on carbonate rock elsewhere have exploded when rising fumes from waste organic solvents dumped into local sinkholes ignited.

Boulders under water weigh less than in the air because the weight of the water partly cancels the weight of the rock. When this lift is removed from a cavern ceiling after pumping lowers the water table, many new sinkholes may open as the caverns collapse. Roads and buildings add weight to cave roofs, helping them to collapse. During design testing of a brand new sewage treatment plant with clean water, the limestone shifted under the added weight, and half a million gallons of water disappeared in about the time it takes for a flush toilet to cycle. Although the loss was not untreated sewage, plugging the hole and stabilizing the ground took another half year.

Many commercial caverns feature boat rides on underground rivers. Clearly, a cavern that does not filter out boats can hardly filter bacteria from the water, as do formations composed of sand or silt. Another complication is that underground flow in carbonate rock does not necessarily follow surface drainage patterns. So impure cave water may cross drainage divides to turn up unexpectedly in wells and springs far removed from any source of contamination.

Limestone terrain is useful, if you know how to use it.

Will this wall retain the boulder?

SEA LEVEL UPS AND DOWNS

Rock textures, structures, minerals, and fossils can tell a geologist whether a sedimentary rock was laid down at, above, or below sea level. And shallow, well-lighted water gives a different signature than does deep, dark water. Algal mats—stromatolites—indicate sunny tidal water and pyrite crystals indicate deep dark water. Rock successions in the geologic column reveal constantly changing sea levels as bed upon bed was laid down. Even today, accurate tidal records show that sea level is rising along the Virginia coast at a rate measured in inches per century. How do we account for these successions of rise and fall of sea level?

The most recent, drastic changes in sea level correspond with Pleistocene ice ages. Winter snows that did not melt captured water evaporated from the sea to form glacial ice on the continents. The amount of water tied up as continental ice dropped sea level world wide by 450 feet—more than once. Melting of Greenland and Antarctic glaciers could raise sea level an additional 200 feet. That would put the Virginia shore near the present Fall Line. And there are other reasons for changing sea level.

Sea floor spreading, for example. Molten basalt rises in the mid-Atlantic ridge to erupt and crystallize as new ocean floor. As this rock cools, it contracts. If the rate of sea floor spreading is relatively fast, say more than five inches per year, the central Atlantic and other spreading centers accumulate a lot of hot crust, effectively making the oceans shallower and causing sea level to rise. If the spreading rate is slow, say less than one inch per year, the crust has time to cool, making the ocean deeper and causing sea level to fall world wide. And movement of the continents can affect sea level, too.

Much of the world's inventory of radioactive elements is concentrated underneath the continents. When a continent sits for tens of millions of years at one spot, the crust and mantle heat up because of radioactive decay, causing the rock to expand and sea level to fall around that continent. If a continent slowly migrates to a low position over cold, recently subducted crust, sea level around that continent will rise.

Limestone ledges above a highway quarry.

If a subduction zone brings an island arc toward a continental margin, that margin may be pulled down in the process causing local lowering of the continent and local rise in sea level. This appears to have occurred in middle Ordovician times when the shallow-water early Paleozoic carbonate bank was succeeded by the deep-water sediments of the Martinsburg formation. That happened while the Taconic orogeny built mountains to the east of the present Blue Ridge and to the north in New England and Maritime Canada.

Growth upward of limestone-producing organisms is much faster than slow subsidence of a tropical continental margin. Given the broad, shallow continental seas of the Paleozoic carbonate banks, rapid growth could build the bank up to sea level and prevent tidal circulation. When this happened, increased salinity in the water could kill off the limestone-producing plants and animals. Slow continued continental subsidence would continue, but the organisms to produce limestone would not be reintroduced onto the bank until water depths increased enough to allow tidal circulation. Repetition of this pattern would be seen in the rock record as regular and cyclic alternations in water depth for the carbonate bank.

Skyline Drive
Shenandoah National Park
Swift Run Gap — Rockfish Gap
39 mi./63 km.

The Swift Run formation is named for outcrops east of Swift Run Gap, in the woods above US 33. The gap cuts through the Swift Run sediments and Catoctin greenstones into Pedlar charnockite, a pyroxene granite. Between the gap and Swift Run Overlook, the drive crosses the Swift Run formation. It forms the slopes above the overlook.

Sharp ridges west of Sandy Bottom Overlook are partly covered by white blocks of Chilhowee sandstone. These treeless slopes are like many developed in that formation. Blocks of Chilhowee sandstone, many complete with *Skolithus* tubes, form the wall around the overlook. *Skolithus* tubes, presumed to be fossil burrows of some wormlike sea-bottom creature, run about one quarter inch in diameter and up to three feet long.

From the Rocky Mount Overlook, Rocky Mount is visible as the main ridge across Hawksbill Creek. The ridge is an anticlinal arch folded in the Chilhowee formation; quartzite talus mantles much of its flank. Hawksbill Creek follows the contact between the Chilhowee sandstone and the Catoctin greenstone. In the roadcut, the massive greenstone is a dark gray-green on unaltered, freshly broken curved surfaces. Slight alteration has bleached the old natural fracture surfaces to a paler color.

Another variant of the Catoctin greenstone crops out at Loft Mountain Overlook. Although some of the rock here is massive greenstone, lenses and pods of sandstone appear in the roadcut across from the overlook. Running water, perhaps only a stream running across a basalt surface, deposited the sand. The rock here is also cut by numerous veins of white quartz and yellowish green epidote. Curved shear surfaces in the rock are coated with white quartz, possibly replacing asbestos.

From Pinefield Gap to the entrance of Loft Mountain campground, Skyline Drive is carved into various Chilhowee sandstones laid down on top of the Catoctin greenstones. The road is back in the greenstone from the service station almost to Big Run Overlook.

The roadcut across from Big Run Overlook exposes Chilhowee sandstone. The rock is moderately coarse-grained, slightly pebbly, red sandstone. If you walk a quarter of a mile north along the road you can see roadcuts in the Chilhowee sandstone, some with feldspar grains, then dark reddish slates, and on to dark reddish basalt with pale feldspar crystals up to nearly an inch long. The feldspar crystals were floating in the basalt lava as it erupted.

From Big Run Overlook to Jarmans Gap, outcrops and roadcuts along Skyline Drive reveal a wide variety of sandstones and related

Azaleas prefer Chilhowee sandstone on the Blue Ridge.

rocks that make up the Chilhowee group of formations. The contact with the underlying Catoctin greenstone is at road level at Browns Gap, but no outcrop is there to reveal it. But you can see that mountain laurel, so common on the greenstone, gives way to azaleas on the sandstone.

At Horsehead Overlook, the direction and steepness of dip of the bedding planes of the Chilhowee sandstone reveals that the entire formation has been folded past the vertical and overturned to the northwest. Thus, it is the northwest overturned limb of an anticlinal arch. Many of the anticlinal folds in the Blue Ridge and Appalachians are similarly overturned.

From Horsehead Overlook to the end of Skyline Drive at Rockfish Gap, views to the west reveal rugged ridges and valleys eroded into Chilhowee sandstone. Many of the slopes are mantled with blocks of the sandstone sliding down from the ridge tops. Even in the valleys, soil is very thin and extremely infertile because the sandstone lacks plant nutrients.

The Shenandoah Valley extends beyond the sandstone ridges all the way to Little North Mountain, which appears in the distance on

Small stream cascading over massive granite.

clear days. The valley is cut into the lower Paleozoic carbonate bank, gently folded into broad synclines and anticlines. Chert in the carbonate formations, along with some sandstone beds, provides weather resistant rocks that stand out as low tree-lined ridges within the valley.

At Moormans River Overlook, the view east is through a gap between Pasture Fence Mountain and Bucks Elbow Mountain. These ridges are capped by Catoctin greenstone, but the Charlottesville reservoir in Moormans River is in the Pedlar granite which underlies the Catoctin greenstone. The North and South forks of Moormans River run parallel to the Blue Ridge in a valley they eroded along a suspected shear zone in the Catoctin greenstone.

From Jarmans Gap to Rockfish Gap (US 250), Skyline Drive is on the Catoctin greenstone. In this stretch, Shenandoah National Park is only as wide as the fences indicate on either side. Beyond the fences the land is privately held and mostly used as mountain pasture.

Beagle Gap Overlook provides a broad view to the east.The distant ridges on the horizon are Brown and Southwestern mountains of the Lovingston massif with another belt of Catoctin greenstone on the far side. Rocks similar to the Pedlar granite underlie these mountains. Between the Lovingston massif and the Pedlar massif—Pedlar granite plus the Catoctin greenstones of the Blue Ridge—the Earth's crust is cut by a massive shear zone, the Rockfish Valley fault. Geologists think this fault connects with other similar faults to the southwest and extends all the way to Georgia.

Rockfish Gap marks the southern end of Skyline Drive and the northern end of the Blue Ridge Parkway. The change is one of administration, not of geology.

Perched greenstone boulder at Greenstone Overlook on the Blue Ridge Parkway.

Blue Ridge Parkway
Rockfish Gap — James River
64 mi./103 km.

The Blue Ridge Parkway begins where Skyline Drive ends at Interstate 64 and US 250 in the Rockfish Gap in the Blue Ridge. This is national forest rather than national park. The rules are a little different and there is no toll. Parking is permitted only in designated areas. *Federal law expressly forbids collecting rock and mineral samples*.

The main rock types exposed in cliffs and roadcuts along the parkway are granite and gneiss of the Precambrian Virginia Blue Ridge basement complex; sandstone and conglomerate of the Precambrian Swift Run formation; greenstone and sandstone of the Precambrian Catoctin formation; sandstone and phyllite of the Cambrian Chilhowee group. Both the Catoctin and Swift Run formations pinch out southward.

The first overlooks to the east provide a broad vista of the Rockfish Valley eroded in the crushed remains of a shear zone. The hills east of the valley were shoved in from possibly 100 miles away and the shear zone is what remains of the zone of transport. This Rockfish Valley fault has been linked with the Fries fault in southwest Virginia, which then connects with the Hayesville fault that runs clear into Alabama.

Because the Rockfish Valley fault dies out to the north in the Blue Ridge anticlinorium, motion along it may be described as the closure of a great scissors once open to the south. This major geologic structure divides the Blue Ridge geologic province and forms the eastern boundary of the Pedlar massif. Rocks east of this fault belong to the Lovingston massif.

Spectacular cliffs in the gray green, fine grained greenstone have been created in some of the roadcuts for the Parkway. Some of these rocks are metamorphosed basalt flows, others are metamorphosed volcanic ash falls. But distinguishing the two types of greenstone is commonly difficult without the aid of a microscope. The United States Forest Service provides a short walking tour of the greenstone next to the valley overlook at milepost 8.8.

Views west provide a panorama of the Shenandoah Valley and its rolling hills. On clear days, the far boundary of the valley can be seen as a discontinuous line of short ridges known collectively as Little North Mountain. Here the North Mountain fault surfaces and separates the double stack of rock strata under the valley from the single stack of strata under the distant Appalachian Mountains.

Greenstone conglomerate across from the Greenstone overlook has been partly replaced by pale green epidote.

South from the overlooks on the west side of the Parkway, the folded and faulted Chilhowee foothills to the Blue Ridge jut into the valley at Big Levels Game Preserve and extend in the distance toward Lexington. Views to the southeast from south of the Rockfish

Mountain farms dot the rolling upland of the Blue Ridge.

Valley reveal rugged mountains eroded into granites and gneisses of the Pedlar massif.

The community of Love straddles the Blue Ridge at milepost 16, and stretches along the road down to Sherando Lake. A few yards beyond this intersection the Swift Run formation appears in the roadcut beside the Parkway. Here it is a mixture of coarse and fine sandstone, phyllite, greenstone, and a big glob of rock altered to green epidote. Individual channel deposits can be recognized in this roadcut.

One of the best panoramas of this part of the Blue Ridge is from the knoll above the Parkway less than a mile north of the Tye River Gap (Virginia 56). To the west and northwest you can see the Shenandoah Valley and the Appalachian Mountains beyond. To the southeast you look down the Tye River valley through the carved granite mountains of the Pedlar massif and across the Rockfish Valley fault to the Lovingston massif beyond. The prominent peak west of the valley is Three Ridges, those to the east are the Priest and Old Maintop. From the next overlook to the north on exceptionally clear days you can see Willis Mountain on the horizon more than 50 miles away, in the Piedmont.

Within this view and somewhat beyond, on the night of August 19-20, 1969, fell one of the greatest downpours of rain ever recorded in the United States. The disorganized remnants of Hurricane Camille, which wrought havoc on the Mississippi Gulf coast two days earlier, encountered another mass of moist air as it passed over the Blue Ridge here. The two air masses combined to drop nearly three feet of rain between 10:30 in the evening and 4 the next morning.

By dawn the raging torrents had claimed 150 lives and damaged this rural area to the extent of $150 million. Houses were destroyed and boulders the size of houses moved. Where bridges were not washed out, the approaches were. Still visible today are the chutes on the mountainsides where water stripped timber and soil to bare granite.

Debate still rages on the question of the expected frequency of such storms. Estimates range from 100 to 1000 years for an average repeat interval, but data now support a period closer to the shorter than to the longer end of the range. Whatever the frequency of these deluges, they must play a major role in eroding these mountains and valleys.

East of the Tye River Gap is the ridge-top community of Montebello with its private and state trout hatcheries. Crabtree Falls, a series of cascades dropping more than 1000 feet in a map distance of 3000 feet, lies beyond Montebello, midway on the Tye River's course down the east side of the Blue Ridge.

The rolling topography for the next 20 miles is underlain by charnockite, pyroxene granite. In some places this charnockite has been recrystallized at much lower temperature than that at which it formed. The result is a very attractive combination of pink feldspar, green epidote, and blue-gray quartz.

Between the world wars, communities and logging camps were connected by the Blue Ridge Mountain Railroad, a trestle of which has been reconstructed at milepost 35. This railroad was a major transportation link for the people who then lived on the Blue Ridge. Abandoned roadcuts for the line can be seen through the trees in several places along the Parkway.

Rebuilt trestle of the mountain railroad that ran along the crest of the Blue Ridge hauling supplies and chestnut logs.

Grenville augen gneiss crops out along the Blue Ridge Parkway south of Otter Creek.

From milepost 39 to the intersection with US 60, the roadcuts expose a variety of rock types of the Chilhowee group—sandstone with conglomeratic lenses and layers of siltstone metamorphosed to phyllite. Views west are interrupted by foothills to the Blue Ridge with their distinctive bare talus slides of Chilhowee sandstone.

Westerly vistas reveal that the Shenandoah Valley is considerably narrower here than farther north. For scale, Interstate 64 cuts across the valley toward the north end of North Mountain. The mesa-like object south of Interstate 64 is actually a double peak, Little House Mountain in front of Big House Mountain, the remnant of a syncline held up by resistant Tuscarora quartzite. The sharp ridge to the south is Short Hills and the double peak to the north is Forge Mountain, a complexly faulted syncline.

From US 60 to Otter Creek facilities, most of the roadcuts are in varieties of sandstone that belong to the Chilhowee group. For about six miles south of the intersection, however, the Parkway diagonally crosses a fault-bounded slice of rock containing both granite and gneiss along with the sandstones. At milepost 48 the rounded outcrop is typical of granite.

At the north end of the Licking Springs Overlook the road is cut into thoroughly sheared granite. A dark diabase dike at the north end provides a good view of an intrusive contact between two rock types. At the middle of the cut, the smooth, slickensides indicate fault movement. Pretty little dendrites of black manganese oxide that look as though they could be fern leaves decorate many of these smooth surfaces.

Lack of plant growth on the face of this cut and the sloping cones of crushed rock beside the road show that it is actively shedding debris. New material falls off with each heavy rain and each cycle of freeze and thaw. Notice how the turf and roots hang over at the top.

For several miles south of US 60, the Snowdon member of the Chilhowee sandstones, a white, pure quartz rock, stands above the trees on the foothills west of the Parkway. Because it is such pure quartz, the rock is extremely resistant to chemical weathering. The Snowden sandstone contains *Skolithus*, a tubular fossil, possibly a preserved burrow.

The view across the valley reveals Short Hills which are held up by the Tuscarora sandstone. North and Forge mountains, also Tuscarora sandstone, are in the distant northwest. In the foreground, the Chilhowee foothills are blanketed with their usual talus of sandstone blocks.

Outcrops north of the tunnel are medium-grained, massive red sandstone beds dipping away from the roadway. South of the tunnel, the road goes into a phyllite. Note the alternating thin red and buff bands and the way the nearly vertical bedding planes intersect the fracture sets to give talus blocks about the size of a deck of playing cards. These features are typical of the Swift Run formation in this area.

Bedding planes of Swift Run formation dip toward the Parkway at the intersection with Virginia 130.

Restored log cabin on the Blue Ridge.

From the tunnel past the Otter Creek campground to the bridge over highway 130, the route weaves back and forth across the contact between the red sandstones of the Chilhowee group and the siltstones and phyllites of the Swift Run formation.

From just south of highway 130 to the James River, the Parkway is cut into an augen gneiss that is progressively less sheared to the south. The augen are pale feldspar crystals as big as a penny set in a matrix of mica and crushed feldspar and quartz. Some zones within this body are free of dark minerals and unsheared.

US 11, Interstate 81
Harrisonburg — Roanoke
114 mi./183 km.

Harrisonburg is near the southwestern end of Massanutten Mountain which forms the core of the Massanutten synclinorium and is held up by the resistant Massanutten sandstone. This sandstone is the remains of a vast sand shoal that accumulated in the Iapetus Ocean during Silurian time, roughly 430 million years ago.

To the west you can catch glimpses between nearby ridges of Little North Mountain in the distance. It, too, is held up by Silurian sandstone, but outside the Massanutten synclinorium it is called the Tuscarora formation. The nearby ridges that block part of the distant views along parts of the routes between Harrisonburg and Staunton are layers of sandstone and chert in the formations of the early Paleozoic carbonate bank.

Visible west from just south of Harrisonburg, is a different kind of conical hill. Mole Hill is the eroded remains of a 50-million-year-old volcano with a plug of olivine basalt for a core. The similar-appearing conical hills, Betsy Bell and Mary Gray, between Staunton and the interstate are different. They stand above the valley because the limestone there contains sandstone and chert.

Between Staunton and Greenville a series of low hills rise in the Shenandoah Valley to the southeast. These are more chert-cored hills in the same limestone formation. In between are the younger Martinsburg shales.

A nepheline syenite dike crops out on the east side of US 11 at the junction between the business and bypass routes north of Staunton.

The Shenandoah Valley abruptly narrows from the southeast because the Chilhowee group of sandstones and phyllites is exposed in a series of folds. From rises in the route, you see these ridges and hills stretching away to the southwest.

The southwestern end of the Massanutten synclinorium is near the interchange with US 11 and US 340. The broad divide between Shenandoah drainage to the northeast and Roanoke drainage to the south is poorly defined because water flows mostly through caverns in the soluble limestone bedrock. East of the route are the headwaters of two rivers named South, one draining into each basin.

Roadcuts in the carbonate rocks are mostly in gray limestone, at many places shot through with white calcite veins. Many of the cuts on the west side of the route are at an angle determined by the eastward dipping bedding planes.

White calcite veins crisscross the dark faces of the cut into limestone of the early Paleozoic carbonate bank near the junction of US 11 and bypass 11 north of Lexington.

Near Lexington, the road follows close to the Pulaski fault. It marks the leading edge of the Pulaski thrust sheet which was shoved onto the Saltville thrust sheet. Cambrian limestones of the upper sheet overlie younger Ordovician limestone of the lower sheet. Lexington is on the lower sheet.

Hills east of Lexington are more of the Chilhowee foothills to the Blue Ridge. The double peak northwest of town is Forge Mountain. What appears at first glance to be a flat-topped mesa west of Lexington resolves itself on closer inspection into Big and Little House mountains. The short, sharp ridge south of the Houses is Short Hills and south of that the Knob may be visible. All these topographic highs contain resistant Tuscarora sandstone.

From just south of Natural Bridge to just north of Buchanan, US 11 and Interstate 84 are together. This stretch of road is carved into a slice of early Paleozoic limestone caught within the Pulaski fault—above the Saltville sheet but below the Pulaski sheet. Such slices, sometimes called "horses," are common complications in the thrust faults of the Appalachians.

Natural Bridge is what remains of a cavern system here. Less than a mile away are caverns that have not yet collapsed.

Purgatory Mountain is the prominent ridge northwest of the route at Buchanan. This ridge is an anticline folded into the Tuscarora sandstone. The same anticline also folds the Pulaski thrust fault. The surface trace of the fault wraps around the nose of the anticline and doubles back to the north for 9 miles on the east side of the James River. It then resumes its southwestward direction on the northwest side of the Fincastle Valley. At Buchanan, as at many other places, the sandstone quarries are in their own way as spectacular as the natural hills carved by running water in the Chilhowee sandstones.

Some of the beds in the Chilhowee group are iron rich, in some places pure hematite. From 1876 until the Great Depression these beds were mined for iron ore, some of which was brought down to the valley below by aerial tramways. Some of the mine entries are still open, but hidden in the second growth forest.

Tinker Mountain, southwest of Troutville, defines the southwestern end of the Fincastle Valley. The mountain is a complex synclinal structure chopped off on its northeast end and southeast side by the Pulaski fault. At Roanoke, the Blue Ridge fault swings back to the east, opening up the Roanoke Valley. An isolated hill in the north end of this valley is the result of another fold in the fault which was eroded through revealing younger rocks below the thrust fault.

US 60, Interstate 64
Amherst — West Virginia state line
92 mi./148 km.

The route begins on the eastern side of the Blue Ridge province. Bedrock here is the Precambrian Stage Road layered gneiss of the

Pedlar charnockite is sheared almost beyond recognition in the Rockfish Valley fault zone where it crosses US 60 west of Amherst.

Sheared Pedlar charnockite is cut by a dike of anorthosite just west of the Blue Ridge Parkway along US 60.

Lovingston massif, Grenville rocks. A sheared zone in this rock is exposed in a roadcut about seven miles west of the circle in Amherst. About nine miles west of Amherst, where US 60 drops down into the valley of the Buffalo River, you cross the Rockfish Valley fault. Pedlar charnockite, some of it sheared almost beyond recognition, crops out in roadcuts there.

Note the size of the Buffalo River just west of the shear zone, and the great breadth of its valley floor. Like many streams draining the eastern slopes of the Blue Ridge in central Virginia, the Buffalo River is here underfit; it has a flood plain that is too big for the size of the river. Geologists do not yet agree on what that means.

A slice of Chilhowee sandstones is faulted into position along the upper reaches of the Buffalo River. You can see dikes of anorthosite along the route, notably where the highway climbs out of the Buffalo River to the west. In the Pedlar River valley, granite dikes and quartz veins intrude granulite gneiss.

The crest of the Blue Ridge is underlain by granites of the Pedlar massif. One of the best "exposures" of the Pedlar charnockites and their variety are in the abutments of the Blue Ridge Parkway bridge over US 60. West of the Blue Ridge are the Paleozoic rocks of the Valley and Ridge geologic province.

The turnout on the Parkway just south of the Parkway Bridge reveals excellent vistas of the Shenandoah Valley and the ridges beyond. Most of the knobs and ridges in the valley are zones of chert and sandstone in the early Paleozoic carbonate bank. The steep, round-topped hill in the middle ground is the nearest of a pair—Big House and Little House—all that remains of an infold of Tuscarora sandstone in the trough of a syncline. Sharp foothills of the Blue Ridge in the foreground are in Chilhowee sandstones with their typical talus slides. Between the Parkway and the Maury River, the route passes into younger sandstones of the Chilhowee group.

The rolling landscape of the Shenandoah Valley is the product of chemical erosion of limestone and dolomite formations of the early Paleozoic carbonate bank. In some places streams disappear into sinkholes dissolved in the limestone. Limestone cliffs rise above the Maury River and dark gray limestone cut by innumerable white calcite veins appears in the roadcuts near the junction of Interstate 64 with Interstate 81.

Two main thrust faults shoved older formations from the southeast on top of younger ones to the northwest. Near the interchange between US 60 and Interstate 81, the Pulaski thrust fault put Cambrian limestone formations on top of younger Ordovician ones. Near the westernmost Lexington exit on Interstate 64, the North Mountain fault created the same arrangement.

Between Lexington and North Mountain, the road climbs up the geologic section through Ordovician rocks to Silurian Tuscarora sandstone at the crest. Simpson Creek, a tributary to the Cowpasture River, follows a synclinal valley in easily eroded Devonian shales and limestones, clearly visible in roadcuts along the northwest side of North Mountain. Water behind these bedding-plane cuts continually topples the shales down toward the roadway. Pipes sunk into these slopes attempt to relieve this groundwater pressure.

Northwest of North Mountain, Brushy and Mill mountains are resistant Tuscarora sandstone on the opposite flanks of an anticline. A spine of bare sandstone rises above the trees on the east flank of Brushy Mountain. Without the Tuscarora sandstone, there would be far less scenery in this province.

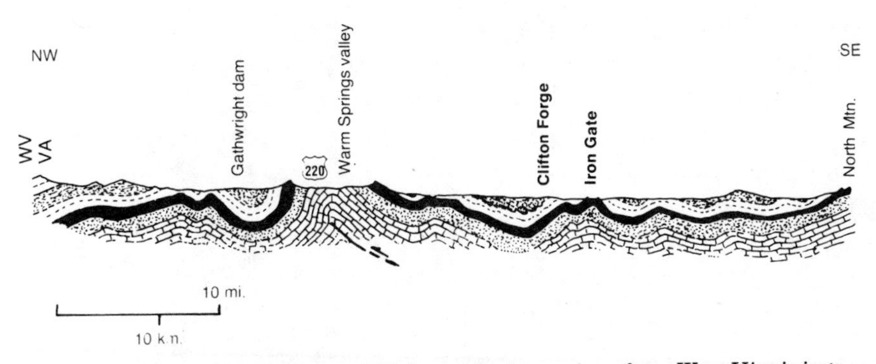

Cross section across Valley and Ridge geologic province from West Virginia to North Mountain southwest of Lexington.

Two tall brick smokestacks are the remains of Longdale Furnace, an iron foundry that operated at the south end of Mill Mountain from the late nineteenth to early twentieth century. Silurian sandstones crop out in the roadcuts and on the mountainsides in this area.

From the Cowpasture River to Covington on the Jackson River, the route follows approximately the boundary between the central and southern Appalachian Mountains. Central Appalachian ridges trend from northeast to southwest. In the southern Appalachians, the trend of the ridges becomes more easterly.

Geologists don't agree on the significance of this bend in the Appalachians. One clue may lie in the similarity of the structures between Clifton Forge and Covington. Both are anticlines, but they are not aligned. If they were once aligned, as some gelogists believe, the Rich Patch Mountains have moved northeast with respect to the Warm Springs anticline. According to this interpretation, the line between the central and southern Appalachians is a fault that shifted the central Appalachians nearly 10 miles to the west.

East from Clifton Forge, you can see the gorge of the Jackson River to the south and the huge exposed anticline of Eagle Rock sandstone is arched above the river at Rainbow Gorge. As is typical of Valley and Ridge anticlines, the southeastern flank of this one dips gently whereas the northwest side is overturned. South of Clifton Forge, the Jackson River joins the Cowpasture River to become the James River.

Devonian shale has been sheared beyond belief in this roadcut on old US 60 east of Clifton Forge. This shearing may be the result of faults nearby or below the outcrop.

Jackson River carved Rainbow Gorge through the Richpatch anticline. Typical of Appalachian anticlines, the northwest limb (left) is steeper than the southeast one (right). Photo by Thomas M. Gathright, II; Courtesy of the Virginia Division of Mineral Resources.

Because the Devonian shales in this valley are weak compared to the sandstone and limestones that hold up the anticlinal ridges, they undergo intense deformation as they are squeezed. Many roadcuts around Clifton Forge expose thoroughly churned shales. Stopping on the interstate is illegal, but you can inspect these shales on business US 60 in Booker T. Washington Park and along old US 60 just east of the railroad overpass east of the interchange with Interstate 64. Concretions mineralized with calcite and pyrite weather out of the cliff at the west end of the park. The roadcut visible from the interstate contains cavities lined with tiny crystals of calcite, quartz, and gypsum.

Engineers building the roadway through these valleys used shale from the cuts to fill the valleys. These shales contain pyrite, which oxidizes to sulfuric acid, and calcite, which dissolves in acid. This chemical reaction permits the roadbed to compact unevenly, giving you bumpy pavement to contend with.

Between Clifton Forge and Covington, the route follows a shale valley cut by the Jackson River between the Warm Springs and Rich Patch anticlines. The Devonian Helderberg limestone rises east of Covington, forming prominent cliffs where the route crosses the river. Just east of the bridge, watch for a small cave in the cliff across the river.

Once a highway bridge, Humpback Bridge now spans Dunlop Creek for picnickers near Covington.

West of Covington, the route crosses a series of open folds in Devonian strata. At the state line, the roadcut follows the originally horizontal bedding planes of the Devonian Chemung formation. Ripple marks made in shallow water some 350 million years ago are obvious. To the west the strata become horizontal as you pass from the Valley and Ridge to the Appalachian Plateaus province in West Virginia.

US 15
Culpeper — James River
64 mi./103 km.

Between Culpeper and Madison Mills, the route crosses flat ground in the Culpeper basin. Sedimentary rocks here were deposited during Triassic and Jurassic time, as the Atlantic Ocean was beginning to open. Low hills to the west are cut in Catoctin greenstone just across the basin's western border fault.

Orange is on Catoctin greenstone off the northeast end of where they hold up Southwest Mountains. Between Orange and Gordonsville the route parallels the southeastern edge of the greenstone belt. Between Gordonsville and Palmyra it is on weathered schists and phyllites of the Candler formation, except around Boswells Tavern where it crosses a small intrusion of dark hornblende gabbro.

At Palmyra, the route crosses onto the Chopawamsic formation on the northwest flank of the complex downfold of the Arvonia synclinorium. This formation consists of light- and dark-colored volcanic rocks interbedded with quartzite and phyllite. It probably accumulated during Cambrian time.

Above the Chopawamsic formation is the Arvonia formation. It is in the trough of the Arvonia synclinorium. At Arvonia, the formation is slate, which contains enough quartz to make it one of the hardest quarried anywhere. Rare fossils date this formation to Ordovician time.

Bremo Bluffs on the James River owe their existence to a weather-resistant sandstone in the Arvonia formation. The synclinal downfold is clearly visible in the outcrops along the James River under the highway bridge. This is the easternmost wrinkle of the Arvonia synclinorium.

Virginia 6
Richmond — US 29
88 mi./142 km.

The route begins at the Henrico County line in the Piedmont province. Quarries here are dug into the Mississippian Petersburg granite. In recent excavations, you can see that it weathers to a pale buff soil.

Between the county line and Manakin, Virginia 6 crosses the Richmond basin which developed and filled with sediments in Triassic and Jurassic time, while the Atlantic Ocean was beginning to open. Most of the sediments came from a source to the west. Lakes flooded the basin at times and peat accumulated in swamps. That peat is now coal and some of it was mined in the last century. Sediments of the basin have also been explored for oil and gas.

Between Manakin and Sabot, the route crosses the western border fault of the Richmond basin, the Hylas fault zone. This zone, 35 miles long, disappears under the Coastal Plain sediments to the northeast. It probably extends south all the way into North Carolina where similar faults are recognized, but geologic mapping in Virginia has not yet revealed to which of the North Carolina faults it connects.

West of the Hylas fault zone are the rock formations of the State Farm gneiss dome. These are Grenville rocks that formed during Precambrian time, about 1200 million years ago. The Blue Ridge basement complex some 50 miles to the west is the same age, with younger rock in between.

The State Farm granite gneiss makes an interesting bulls-eye on the geologic map. It is surrounded by the Sabot amphibolite which is in turn mantled by the Maidens gneiss, except where that is cut out by the Hylas fault zone. Geologists are not certain how much of this surrounding rock is the same age as the State Farm gneiss. One argument favors including everything east of the Spotsylvania lineament in a Goochland terrane that extends from Fredericksberg to the North Carolina state line. Another limits the Grenville terrane to the State Farm gneiss, proper.

You get good views of the flood plain of the James River west of Sabot and east of Columbia. Although this flood plain may seem excessively wide, every few decades heavy rain sends water flowing across all of it.

At Columbia, the Columbia granite is nicely exposed in a small quarry on the north side of Virginia 6. This granite is part of the Hatcher complex of igneous rocks. Above the Hatcher complex are the metamorphosed volcanic rocks of the Chopawamsic formation. Intrusions of the Hatcher complex appear to penetrate the overlying Chopawamsic formation, so they may all come from the same source. Perhaps the volcanic rocks erupted from the masses of magma that recrystallized at depth to become the granitic rocks of the Hatcher complex.

Columbia lies at the southeast end of the Columbia syncline, a downfold that has slates of the Arvonia formation in its core. Sparse fossils indicate that the Arvonia formation was deposited during Ordovician time. Age dates and cross-cutting relationships lead geologists to conclude that the volcanic rocks of the Chopawamsic formation erupted during Cambrian time.

Quartz veins in the schists, gneisses and quartzites of the Columbia syncline were mined for gold from the early 1800s until the end of World War II. Native gold was the chief ore mineral, but some gold was recovered from sulfide minerals such as pyrite. Although some ore shoots contained more than 100 grams of gold per ton, most had only a few grams per ton. Gold distribution was sporadic. Quartz veins changed from rich ore to barren rock in a matter of feet.

Between Columbia and Fork Union, the route crosses igneous rocks of the Hatcher complex. Fork Union lies near the axis of the Arvonia syncline, which also contains Ordovician Arvonia slates. If the Arvonia and the Columbia synclines are related, they represent a fold belt that stretches more than 100 miles between Danville and

Soapstone quarries at Schuyler fill with water after they are abandoned.

Blocks of quarries soapstone lie discarded because fractures and calcite veins make them useless for further slabbing.

Fredericksburg. Some geologists regard this fold belt as a suspect terrane, which means they think it may be a piece of crust that did not form where we see it, but moved in from elsewhere.

Although there is no record of commercial production, manganese minerals occur in a quartz vein north of Fork Union. The primary mineral is rhodonite, a shocking pink manganese silicate, which weathers to black dendrites of manganese oxide. The dendrites make little filigrees on the white quartz.

From Columbia to Scottsville, where the road again drops down into the valley of the James River, the route crosses typically rolling Piedmont terrain carved into metamorphosed sedimentary rocks of the Lynchburg group. East of Scottsville the landscape flattens where the route crosses onto the Scottsville basin with its fill of sedimentary rocks laid down during Triassic and Jurassic time. The landscape becomes distinctly more rolling west of Ballinger Creek where the route climbs out of the Scottsville basin and back onto the more resistant rocks of the Lynchburg group.

Soapstone blocks were used to build the abutments for the now abandoned railroad bridge over the turn-off for Schuyler, the location of television's Crabtree Falls. Operating and abandoned soapstone quarries are in Schuyler, center of the country's largest soapstone production belt.

Just east of the junction of Virginia 6 with US 29 and in the pastures north of the road are bare pavements of the Rockfish Valley granite. Magma invaded Grenville age rocks of the Lovingston massif about 700 million years ago to form this small intrusion.

US 29
Culpeper — Amherst
93 mi./150 km.

From the western border fault of the Culpeper basin in Culpeper, the route swings westward across the trend of the eastern belt of Catoctin greenstone, the eastern edge of the Blue Ridge province. About five miles west of Culpeper, the route passes onto rocks of the Lynchburg group of schists and gneisses. Some of the schists contain graphite. Thoroughfare Mountain, southeast of Brightwood, is one flank of a fold in the metamorphosed sedimentary rocks of the Lynchburg group.

Between Brightwood and Charlottesville, US 29 follows the trend of the rock layers approximately 15 miles southeast of the Blue Ridge, which is distinctly visible on the northwest horizon. About two miles southwest of Brightwood, the road crosses the contact where the sediments that became the Lynchburg group abut the Robertson River granite.

The Robertson River granite invaded the crystalline core of the Blue Ridge anticlinorium in late Precambrian time. Although less than five miles wide, its outcrop belt extends for about 70 miles in a northeast-southwest direction. It crops out in the fields on both sides of the highway near the Robertson River and it is used in the foundations of some buildings in the town of Madison.

The near ridge on the northwest side of Madison is Gaar Mountain. Part of it is eroded in slivers of Old Rag granite, the main body of which holds up Old Rag Mountain some dozen miles to the north. The Old Rag granite is distinctive for its thumb-sized feldspar crystals that weather out as nubs on natural exposures.

Southeast of Ruckersville are the Southwest Mountains, the

southwestern extension of the Catoctin greenstone belt that begins at Catoctin Mountain at the Potomac River. They are the southeast flank of the Blue Ridge geologic province.

About a mile and a half north of the bridge over the North Fork of the Rivanna River, the route passes onto younger rocks of the Lynchburg group. A gneiss outcrop is on the east side of the highway just south of that bridge. The cut at the north end of the Barracks Road shopping center in Charlottesville provides another good view of these formations.

At Charlottesville, US 29 passes onto older rocks of the Lovingston massif, part of the Blue Ridge province. Somewhere in this general area the geologic character of the Blue Ridge province changes. In northern Virginia it is a distinct complex anticline or anticlinorium with ridges of Catoctin greenstone on both sides except where the western border fault of the Culpeper basin chopped it off.

The north end of the Rockfish Valley fault dies out somewhere south of Old Rag Mountain. It is a thrust fault that shoved rocks of the Lovingston massif onto the Pedlar massif. Displacement along this fault increases to the southwest in a scissors fashion and may be several dozen miles in the Rockfish Valley west of Charlottesville. In southern Virginia, this fault is known as the Fries fault. Its horizontal displacement there is even greater than in the Charlottesville area. Geologists connect it with the Hayesville thrust fault in Georgia. That makes it nearly 1000 miles long.

The route stays on rocks of the Lovingston massif from Charlottesville to midway between Amherst and Lynchburg. The massif contains both massive granite and streaky gneiss with the same mineral composition. The roadcut just south of the bridge over the Rockfish River at the intersection with Virginia 6 west provides a good exposure.

Fold in Tonoloway limestone near Blue Grass.

US 220
West Virginia state line — Covington
61 mi./98 km.

The route ascends the South Fork of the Potomac River toward its headwaters near Monterey. The stream eroded its valley along the trough of a syncline. Devonian limestone and shale are in the core of the fold. Ridges of older Tuscarora sandstone rise on either side of the valley. A spectacular view of vertical ledges of sandstone on the flank of the fold is visible from the schoolyard in Blue Grass, 2 miles west of the highway near the state line. Roadcuts along that side road reveal an array of small folds and kink bands in the Tonoloway limestone, a Silurian formation.

Between the state line and Monterey, cliffs of Devonian limestone rise east of the stream. The layers of rock are very close to horizontal indicating that they are in the flat bottom of the syncline. Flat bottomed synclines bounded by overturned anticlines are typical structures of the Valley and Ridge geologic province in northern Virginia.

Just south of Monterey, Trimble Knob rises 200 feet from the valley floor. Blocks of basalt cap the knob indicating that it is a volcanic plug, the eroded ruin of an extinct volcano. Other basalt and andesite intrusions dot this valley and surrounding hills, but none is so distinct topographically. Dark gray areas in the fields and small valley hills mark the locations of less resistant volcanic intrusions.

Between Monterey and the Bath County line, the route descends the Jackson River in the same syncline. Back Creek Mountain on the west and Jack Mountain on the east are ridges of Tuscarora sandstone, the opposite flanks of the fold. Limestone outcrops abound in this valley, especially west of the highway. Solution by ground water has etched out the bedding planes in many places.

Near the county line, the highway climbs out of the syncline and crosses Jack Mountain to the valley eroded through the arch of the Warm Springs anticline. Tuscarora sandstone forms the ridges on both sides of the valley and cherty limestone of the early Paleozoic carbonate bank makes up the valley floor. Knobs rise in the valley where the chert is most abundant. Grassy slopes higher on the valley wall mark exposures of the Reedsville shale and siltstone.

Typically, the rocks on the west flank of the anticline are more steeply tilted than those on the east. That makes them easier to erode, hence lower in elevation. Gaps in the west wall mark the locations of transverse fracture zones. Those zones also cut the east wall, but can be detected there only on aerial photographs.

The Bath County Pumped Storage Facility is carved into a Devonian shale valley. During periods of low demand, water is pumped from the lower to the upper reservoir where its fall can generate electricity during periods of peak consumption.

Warm springs in this valley gave Bath County its name. Some of them near the center of the valley occur at the intersections of fracture zones and early Paleozoic limestones. Those on the sides of the valley flow where fractures cut the younger limestones on the flank of the anticlinal arch. South of Healing Springs the highway is unstable because ground water has dissolved caverns in the underlying limestone. Cavern collapse leads to highway collapse and the need to fill and repave the road.

Temperatures in the thermal springs are closely tied to rock stratigraphy and structure. The warmest springs flow from the oldest rocks, those lowest in the geologic section. Those rocks are exposed at the center of the fold where the springs reach a temperature of about 105 ° F. Younger Silurian and Devonian limestones on the flanks of the anticlines are substantially cooler because those formations don't carry the water so deep in the earth.

The Warm Springs airport is on Tuscarora sandstone high atop Warm Springs Mountain, the east flank of the anticline. The contact between the Tuscarora and the underlying red Juniata sandstone crops out in the cut for the access road just off the end of the runway.

The first ridge northwest of the airport is Little Mountain, the west limb of the Warm Springs anticline. The second ridge is Back Creek Mountain, another anticline, with the axis near the crest of the ridge. Pine trees mark flatirons of Devonian Oriskany sandstones, great slabs of rock carved on the flank of the mountain. Poplars grow on the underlying Helderberg limestones behind the flatirons. The third ridge is Allegheny Mountain, a syncline topped by Mississippian Pocono sandstones. Ridges underlain by the crest of an intact anticline tend to have even, skyline profiles. Ridges on sandstones flanking anticlines opened by erosion along the crest tend to have very uneven crests.

Early Paleozoic limestones appear in roadcuts and fields for 10 miles south of Hot Springs. The hills and ridges in this valley are honeycombed with caves of all dimensions, some large enough to house summertime balls in the nineteenth century. Land owners, however, found trespassers to be troublesome, so many of the cave entrances have been dynamited or filled with rusted-out automobile bodies.

The south end of the valley drains through a gap above the hamlet of Falling Springs. This is the location of yet another shear zone in the rock. Its effects can be seen in the fractured and broken Tuscarora sandstone in the roadcut across from the falls. The stream that goes over the falls rises from a thermal spring about a mile north. The water is supersaturated with carbonate where it flows under the road

Warm water supersaturated with calcite tumbles over Tuscarora sandstone above the village of Falling Springs.

toward the falls. Leaves and twigs at the bottom of the falls are often coated with white crusts of calcite, especially in late summer and early autumn.

A few yards down the side road leading to Falling Springs, travertine crops out in the roadcut. It looks just like the travertine forming today at the base of the falls. Its elevation above the present falls indicates that this gap in the ridge was higher than it is now when the travertine formed. At one time the travertine farther down the valley was mined for its lime content.

From this gap in the Warm Springs anticline south to Covington, the road follows the west flank and nose of the anticline over a roadbed cut in steeply tilted Devonian shale. Water in sandstones and limestones beneath the shale builds up an hydraulic head behind and below the roadcuts causing landslides, which make the road surface uneven.

Some of the Devonian Oriskany sandstone and underlying limestone has been replaced by limonite, a mixture of iron oxides. This limonite is rich in nickel and manganese, impurities that made iron that was tough and in high demand for Confederate military ordnance. The limonite also contains zinc, which collected in the flues. The furnace was shut down from time to time so the operators could clean zinc from the flues.

Rising water destabilized the mountain slopes above Lake Moomaw causing entire slopes to slump into the lake.

Weathering of the iron sulfide mineral pyrite provided acid solutions capable of keeping dissolved iron in solution. When this acid ground water reacted with limestone, the limestone dissolved leaving cavities in which limonite collected. Some of the overlying sandstone was also replaced by limonite. Most of the old mine workings are caved in and overgrown, but their locations are shown on topographic maps.

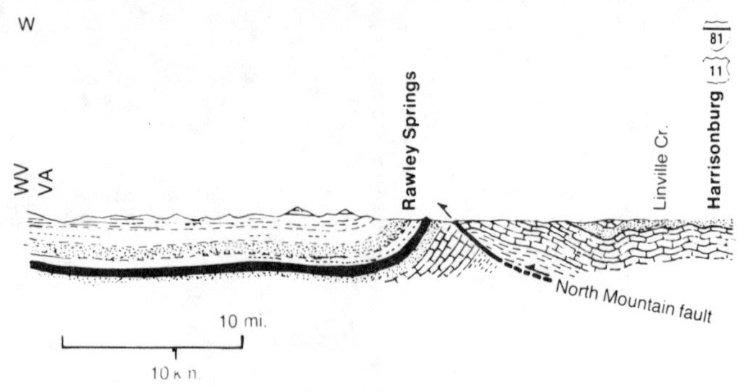

Cross section of Valley and Ridge geologic province

US 33
Richmond — West Virginia state line
136 mi./219 km.

Northeast of Richmond, the topography is very flat where the road crosses terrain eroded into the Mississippian Petersburg granite. Soil and saprolite developed on this granite is pale to buff colored in fresh excavations. Outcrops in Maidens gneiss are northeast of the bridge over the South Anna River. Diabase dikes reveal their presence by short, sharp rises in the road with deep red soil in the ditches alongside the road.

At Montpelier, the route crosses the north end of the State Farm dome. This Precambrian structure may be an isolated piece of billion-year-old continental crust that drifted off the Blue Ridge when the Iapetus Ocean opened. Or it may be the core of a much larger terrane extending east of the Spotsylvania lineament from the Po River metamorphic complex near Fredericksburg to the Raleigh belt in North Carolina. The feldspar quarry south of town is in an anorthosite body within the State Farm dome.

Northeast of the junction with US 522 at Cuckoo, the topography becomes typical rolling Piedmont as the route crosses onto schists of the Arvonia formation. Two miles southeast of Louisa, a side road, highway 767, cuts into recognizable but rotten, deeply weathered schist. Southwest of here, quartz veins in the rocks of the Columbia syncline were mined for gold.

From Louisa, you see the Southwest Mountains, the southern extension of Catoctin Mountain, the eastern flank of the Blue Ridge

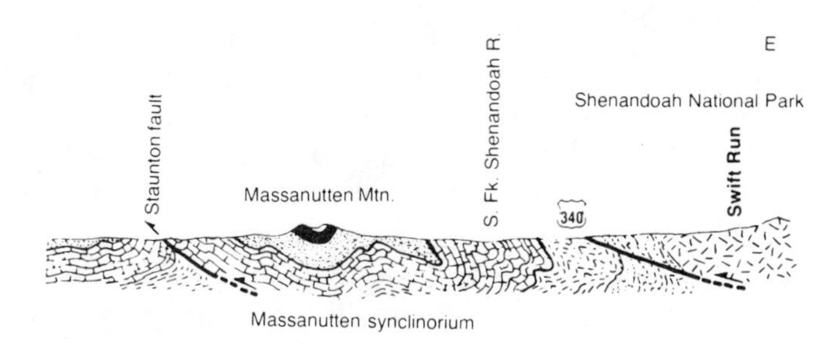

parallel to US 33 from West Virginia to the Blue Ridge.

province. Between Gordonsville and Ruckersville the route crosses a corner of a Triassic basin that was once continuous with the Culpeper basin. Ruckersville is on weathered Robertson River granite.

Between Ruckersville and Stanardsville, the route crosses the Batesville graben, a block of basement that dropped along faults as the Iapetus Ocean opened in Eocambrian time. There are no exposures of basin fill along the major highways. West of Stanardsville, the route crosses the Rockfish Valley fault, which moved the Lovingston massif of the Blue Ridge over the Pedlar charnockite.

US 33 passes under Skyline Drive on Pedlar charnockite at Swift Run Gap. About one half mile west of the underpass the road passes onto Catoctin greenstone, which was deposited on the eroded charnockite surface as a basaltic lava flow about 650 million years ago. As the road goes down from the top of the Blue Ridge it progresses up the geologic section to younger and younger rock formations.

In the valley of West Swift Run the road goes from the top of the Catoctin greenstone into the lower formations of the Chilhowee sandstones. Manganese and iron were mined from the Chilhowee sandstones both north and south of here. About a mile east of US 340 in Elkton, the route crosses onto the limestone formations of the early Paleozoic carbonate bank that underlie the Shenandoah Valley.

The Shenandoah Valley is eroded into limestone and dolomite formations of the early Paleozoic carbonate bank, formations deposited under conditions resembling those found today off the south coast of Yucatan and the east coast of Australia. In Pennsylvanian time, crustal movements shoved those formations northwest, rumpling them into broad, open folds cut by thrust faults.

Massanutten Mountain, a ridge of resistant Massanutten sandstone, divides the Shenandoah Valley in two from here to Front Royal. The route wraps around its southern end. The structure of the Shenandoah Valley is a synclinorium—that is a broad syncline with smaller folds in it. Massanutten Mountain is the core of the synclinorium. South of the mountain, where the massive sandstone has eroded away, or was not deposited, the core is in the underlying Martinsburg shale, which is easily eroded and leaves no topographic expression. The road passes onto shale west of the South Fork of the Shenandoah River.

Beyond the nose of the mountain, where the beds in cliffs can be seen dipping toward the ridge, the road again starts down the geologic section and back onto the early Paleozoic carbonate bank at Penn Laird. Between Penn Laird and Interstate 81, the route crosses the north end of the Staunton fault, which thrusts Cambrian limestone from the southeast onto Ordovician limestone. This thrust fault dies out in a small anticline a few miles to the northeast.

Although Harrisonburg is west of Massanutten Mountain, it sits astride the watershed between the North and South forks of the Shenandoah River. The gray stone used in the buildings of James Madison University came from quarries in the Edinburg limestone. It weathers to gray from its original black, hence the trade name, Jamieson black marble.

West of Harrisonburg the skylines are dominated by Massanutten Mountain to the southeast and Shenandoah Mountain in the west. The conical peak rising some 400 feet above the highway to the south is Mole Hill, one of Virginia's two known volcanic plugs. The early Paleozoic carbonate bank is intruded by other volcanic rocks in the form of sills and dikes, but they do not stand out topographically. The latest volcanic activity was during Eocene time, about 50 million years ago.

Mississippian sandstone and siltstone in roadcuts along US 33 near Rawley Springs.

The progression to older rocks at the northwest side of the synclinorium is truncated by the North Mountain thrust fault. It transported Cambrian and Ordovician limestone of the Shenandoah Valley northwestward over the Devonian and younger rocks of Shenandoah Mountain.

Younger rocks along the northwest side of the fault are tipped up by frictional drag. Farther west, these formations flatten out. Because the rocks are lying flat, to climb Shenandoah Mountain is to go up the geologic section through progressively younger strata. Most of the slopes of the mountain are underlain by Devonian shales. The core of the broad syncline contains the Mississippian Pocono sandstone, one of the prominent ridge formers of the western Valley and Ridge province. These red sandstones are exposed in the roadcut at the West Virginia state line.

US 340
Elkton — Greenville
48 mi./77 km.

The southwestern nose of Massanutten Mountain is visible from Elkton on US 33. Although the mountain ends here, the synclinal structure that made it continues southwest for another 30 miles to near the end of US 340 at US 11. In this stretch of the syncline, the core is Martinsburg shale, which is not resistant enough to hold up sharp ridges.

Between Elkton and Waynesboro, US 340 runs on terraces and alluvial fans containing cobbles and boulders of quartzite, granite, and greenstone washed down from the Blue Ridge. Watch for them in many stream bottoms. Ridges and low wooded hills west of the highway between Elkton and Waynesboro mark beds and lenses of sandstone and chert. They stand high because they resist weathering better than the early Paleozoic carbonate rocks that underlie most of the Shenandoah Valley. Sharp hills and ridges east of the highway are eroded in Chilhowee sandstone and siltstone. Around the turn of the last century, more than half the country's manganese came from a single mine at Crimora.

The prominent scar on the hillside west of Waynesboro and visible from much of the downtown area is a quarry in Chilhowee sandstone. Southwest of Waynesboro, US 340 cuts diagonally across folds in the Cambrian to Ordovician carbonate sequence toward Stuarts Draft.

Curtains of calcite drape from the ceiling in this cavern passageway.
Photo courtesy of Shenandoah Caverns.

Rugged hills to the south in the Big Levels Game Preserve are on folded Chilhowee sandstones.

The junction of US 340 with US 11 is in the early Paleozoic carbonate bank sequence near the southern end of the Massanutten synclinorium. To the south the Chilhowee foothills to the Blue Ridge stretch off to the horizon and glimpses of Little North Mountain can be seen to the west.

US 250, Interstate 64
Richmond — West Virginia state line
141 mi./227 km.

Richmond straddles the Fall Line, the western onlap of Coastal Plain sediments onto the crystalline rocks of the Piedmont province. These igneous and metamorphic rocks are weathered to saprolite. In places, it is more than 100 feet deep. Except for a few resistant rocks, such as quartzite, bedrock is exposed only where rivers and streams have stripped away the overlying saprolite. Because these routes follow the high ground between the Anna and James river systems, very little bedrock is exposed along the highways, even in fairly deep roadcuts.

The thin layers of Coastal Plain sediments at Richmond rest on the Petersburg granite, which intruded older Piedmont schists and gneisses in Mississippian time, about 330 million years ago. A quarry exposes rock west of Interstate 295 between Interstate 95 north and Interstate 64 west.

The interchange between Interstate 295 and Interstate 64 west of Richmond is on granite. About a mile to the west, the route passes onto the northern end of the Richmond basin, which dropped along faults as the Atlantic Ocean opened during Triassic and Jurassic time. Here the basin is only about two miles wide. Its western border is the Hylas fault zone, a mass of sheared rock nearly two miles wide and more than 30 miles long. It extends well beyond the basin.

West of the Hylas fault zone, the routes pass onto the State Farm

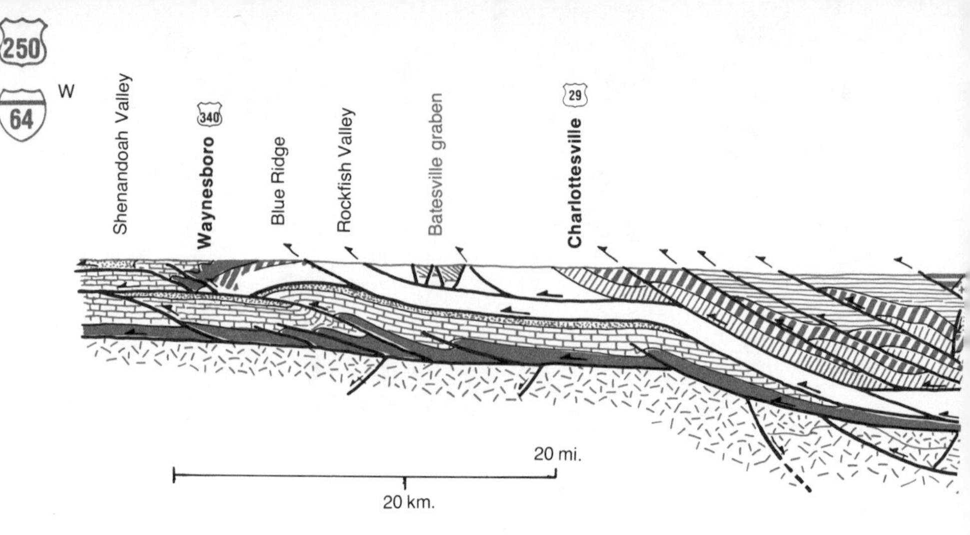

Cross section of Valley and Ridge, Blue Ridge, and Piedmont geologic

gneiss dome, which formed about 1 billion years ago. The Sabot amphibolite, a dark rock full of black amphibole, and the Maidens gneiss wrap around the margins of the dome. At about the US 522 interchange the routes cross a mysterious feature known as the Spotsylvania lineament. This line, which shows up on magnetic maps, may mark an important fault or suture of Paleozoic age.

At about milepost 150 on the Interstate, near the line between Goochland and Louisa counties on US 250, lies the axis of the Columbia synclinorium. It is a fold belt wrinkled into metamorphosed Cambrian and Ordovician sedimentary and volcanic rocks that extends from the James River to Fredericksburg. Geologists relate it to the Arvonia synclinorium, so the fold belt may extend at least as far southwest as Farmville, a total length of nearly 100 miles.

The routes cross the boundary between the Piedmont and Blue Ridge geologic provinces near Shadwell. Watch the roadcuts along

Mafic and felsic rocks of the Hatcher complex crop out along US 250 west of Richmond.

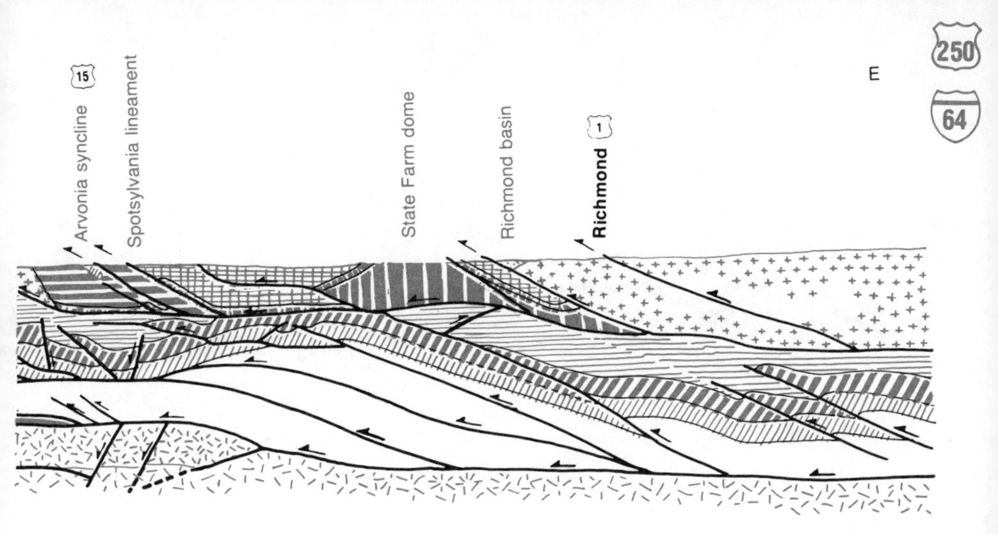

provinces along Interstate 64 from Staunton to Richmond as determined from seismic exploration techniques.

highways east of Charlottesville and in the stone quarry just east of the Shadwell interchange, for metamorphosed sedimentary rocks and Catoctin greenstone, the resistant formations that form the Southwest Mountains, north of the highways.

Charlottesville is on rocks of the Lynchburg group. Although less resistant than the greenstones to the east and the layered gneisses to the west, they are prominently exposed in the interstate roadcuts south of Charlottesville. They give the city its rolling topography.

The boundary between the Catoctin greenstones and the Blue Ridge basement crystalline complex is near the US 29 interchange with Interstate 64 or the US 250 business and bypass interchange. Basement rock here is the Stage Road layered gneiss of the Lovingston massif. Watch for it in the roadcuts on Interstate 64 west of Charlottesville.

Mechums River flows in the Batesville graben. It is a dropped fault block that formed about 600 million years ago and resembles the younger fault block basins that formed in the Piedmont as the Atlantic Ocean was opening during Triassic and Jurassic time. The Batesville graben extends to the northeast a total distance of 60 miles and is mostly less than 2 miles wide. It is probably the same age as the Lynchburg group, but the sedimentary rocks that fill it were deposited in shallower water.

Between Mechums River and the Blue Ridge the routes cross the valley of the Rockfish River. Almost the entire valley is floored by the sheared zone of the Rockfish Valley fault, which separates the Pedlar

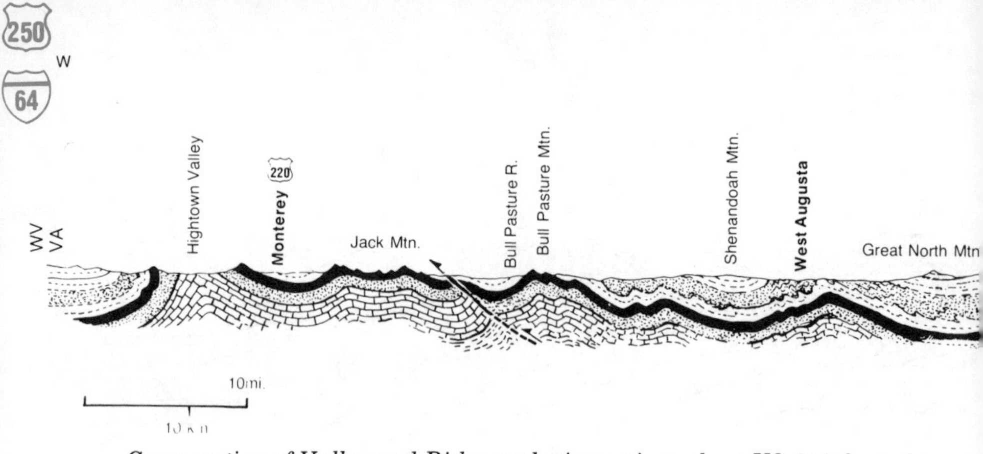

Cross section of Valley and Ridge geologic province along US 250 from the

and Lovingston massifs. The Blue Ridge here is an anticline with its northwestern flank tilted beyond vertical. The dark green rock in the roadcuts is greenstone. The pale green color is the mineral epidote, an alteration product of the greenstone. The greenstone weathers to a rust red soil.

The eastern slopes of the Blue Ridge are underlain by sandstone of the Chilhowee group, but the massive highway interchange in Rockfish Gap is on Catoctin greenstone, phyllite, and dark sandstone. Watch for the sandstone on the north side of US 250 just west of the

Rocks of the early Paleozoic carbonate bank underlie a broad flood plain west of the Blue Ridge near Lyndhurst. The Blue Ridge rises behind craggy foothills in Chilhowee sandstones. Photo by Thomas M. Gathright, II; Courtesy of the Virginia Division of Mineral Resources.

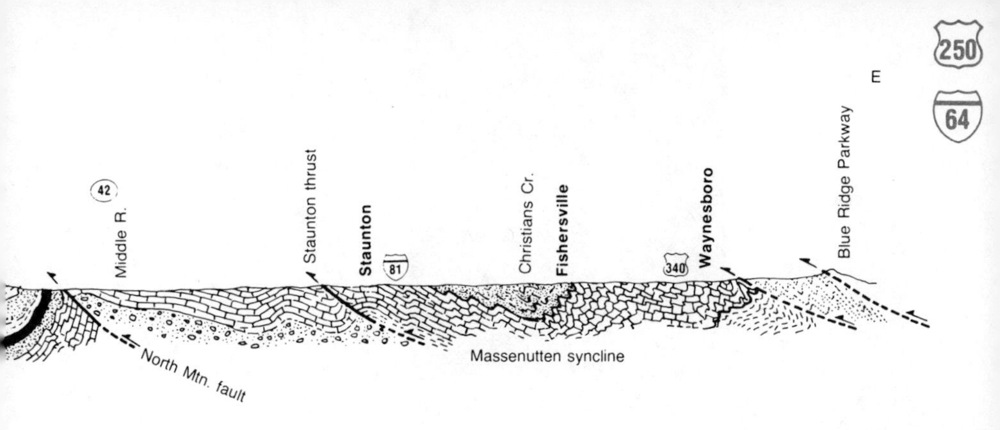

Appalachian Plateaus in West Virginia to the Blue Ridge.

Blue Ridge Parkway and Skyline Drive overpass. The greenstone puts in an appearance across from the Rockfish Valley Overlook on US 250 east of the overpass, or on the overlook off the eastbound lane of the interstate.

The giant blond scar dominating Waynesboro from the ridge east of town is an abandoned quarry, now the city dump, in the Antietam sandstone of the Chilhowee group. The high ground in the middle of Waynesboro is underlain by the Waynesboro formation, part of the early Paleozoic carbonate bank.

The Massanutten syncline, its axis near Christians Creek, lies between Staunton and Waynesboro. The most conspicuous hills in that area are knobs held up by chert or sandstone in the carbonate bank. The center of the syncline is Martinsburg shale, which is exposed in several roadcuts along the railroad tracks. Look closely at some of the roadcuts along US 250 to distinguish between sedimentary layers and the cleavage planes that split the shale.

Between Interstate 81 and US 11, Betsy Bell and Mary Gray, two conical hills of cherty Beekmantown dolomite, dominate the landscape. When geologists discovered that Trimble Knob in Monterey and Mole Hill near Harrisonburg are in fact volcanic plugs, they dragged magnetometers over this pair and every other conical hill in the valley to try to find more magnetic volcanic rock.

The Staunton thrust fault crops out in the roadcut across US 250 from the west end of Gypsy Hill Park in Staunton. Cambrian limestones here are shoved northwest over younger Ordovician limestones. Shattered breccia composed of angular blocks from the size of your thumb to that of your car marks the fault zone. The breccia

Natural Chimneys are all that remain of an enormous cavern system that collapsed to form a gigantic sinkhole. The peaks reach 120 feet above the present surface while drilling has revealed gravels more than 260 feet down. Just above the base of the spires, a six-inch sill of basalt intruded the Conococheague limestone.

contains many calcite veins. You can collect crystal fragments with cleavage surfaces the size of a quarter.

Most of the bedrock underneath the Shenandoah Valley is limestone and dolomite of the early Paleozoic carbonate bank. Between Jennings Gap and Staunton, US 250 crosses the Long Glade syncline and the Middlebrook anticline along with several smaller folds. Outcrops in fields and roadcuts show bedding planes that tilt as much as 40 degrees. Jennings Gap marks the boundary between the Allegheny Mountains and the Shenandoah Valley. Older valley rocks were shoved onto younger Alleghany rocks along the North Mountain fault, which surfaces along the east flank of Little North Mountain. Rocks visible just north of the highway are an isolated slice within the thrust fault. West of the boundary, the rocks tend to be more tightly folded than in the Shenandoah Valley.

West Augusta, on the Calfpasture River, is near the center of a valley eroded through the crest of an anticline. Middle Devonian sedimentary rocks are in the core of the fold. Great North Mountain is a syncline with Devonian red shales and sandstones in the trough of the fold. Shenandoah Mountain, along the Highland-Augusta county line, is another syncline that contains gray and green Devonian shales, siltstones, and sandstones. The axis of the fold is east of the Civil War fortifications that made the defensive slope to the west a particulary steep one to attack. Cowpasture River follows a valley

Limestone is tightly folded in a roadcut one half mile east of the crest of Bullpasture Mountain.

carved in nonresistant beds between an anticline and a syncline.

Bullpasture Mountain is another anticlinal ridge propped up by Tuscarora sandstone. North of the highway on the east flank of the mountain is an abandoned highway quarry where you can collect late Devonian fossils from the Keyser limestone. In addition to crinoid stems, there are brachiopods and trilobites, especially in the outcrops fairly high above the road and just below the trees.

Monterey on US 220 is at the very headwaters of the South Branch of the Potomac River, with the headwaters of the Jackson River less than half a mile south of town. These two rivers cut their valleys into a syncline that contains Devonian shale in its center. Ridges of resistant Tuscarora sandstone stand along the opposite flanks.

Many small basalt and andesite plugs and dikes intrude the shales of this valley. Some have been dated as 50 million years old, Eocene. Although most of them have little topographic expression, Trimble Knob on the south side of Monterey is a basaltic plug that rises approximately 200 feet above the valley floor. Limestone rubble in some of the intrusions suggests to geologists that these volcanic eruptions were explosive enough to blast rubble and lava high into the Eocene sky.

Between Monterey and Hightown the route crosses Devonian shales and Ordovician rocks of the lower Paleozoic carbonate bank. Monterey Mountain in between the two is held up by resistant Tuscarora sandstone, as is Little Mountain on the northwest side of the Hightown Valley anticline. To the north, along the crest of Monterey

Mountain, this formation rises vertically above the trees as the Devil's Backbone.

Conical knobs dot the floor of the anticlinal valley like so many volcanic plugs, which they are not. They are chert-rich parts of the limestones. Eocene andesite and basalt dikes can be seen in the walls and rubble of a small quarry 2 miles north of Hightown, but the small igneous intrusions of the valley don't make significant hills.

Between Hightown and the West Virginia line the road climbs up the geologic section from Ordovician rocks into younger Devonian shales. Back Creek follows these easily eroded shales along the flank of the anticline. Devonian shales in roadcuts on the east side of Back Creek are overturned to the northwest. On the west side of the valley the dips become progressively gentler as you climb the bluff. Watch the roadcuts. This flattening marks the boundary between the Valley and Ridge province to the southeast and the Appalachian Plateaus province to the northwest. To the east the parallel, even topped ridges of the Valley and Ridge dominate the skyline. To the west are the more irregular hills of the Appalachian Plateaus. Compare the pattern of the straight valley cut by Back Creek in the flank of the anticline with the undisciplined dendritic pattern of the next valley to the west. A map of all the little streams feeding into it would resemble a bare hardwood tree against a winter sky.

Trimble Knob in Monterey has a basalt plug approximately 50 million years old in its core.

US 522
Culpeper — James River
69 mi./111 km.

South of Culpeper, the route crosses the Culpeper basin and skirts the north end of the Southwest Mountains, held up by Catoctin greenstone. It passes onto the metamorphic rocks of the Piedmont province just south of the bridge over the Rapidan River. The landscape is distinctly more rolling in the Piedmont than in the Culpeper basin with its fill of weak sedimentary rocks.

At the North Ann River, the route crosses a granite intrusion, perhaps part of the Hatcher igneous complex. Mineral is on the Chopawamsic metamorphosed volcanic rocks of the Columbia syncline. Gold mines that operated in quartz veins that cut the rock southwest of Mineral probably gave the town its name. Although most of the production was native gold, auriferous, or gold-bearing fool's gold, pyrite was also mined. Profits came both from the gold and from sulfuric acid made from the pyrite. Extremely pure native sulphur mined along the Gulf Coast now provides better raw material for sulfuric acid.

Between Gumb Spring and the James River, the route crosses the Maidens gneiss, which surrounds the State Farm dome to the east. The State Farm gneiss is intruded by billion-year-old Grenville granite, so the Maidens gneiss may be even older. If this bedrock is a fragment of a continent, the Columbia syncline to the west may be the crumpled remains of the Iapetus Ocean floor that lay between the State Farm gneiss and the North American continent in Paleozoic time.

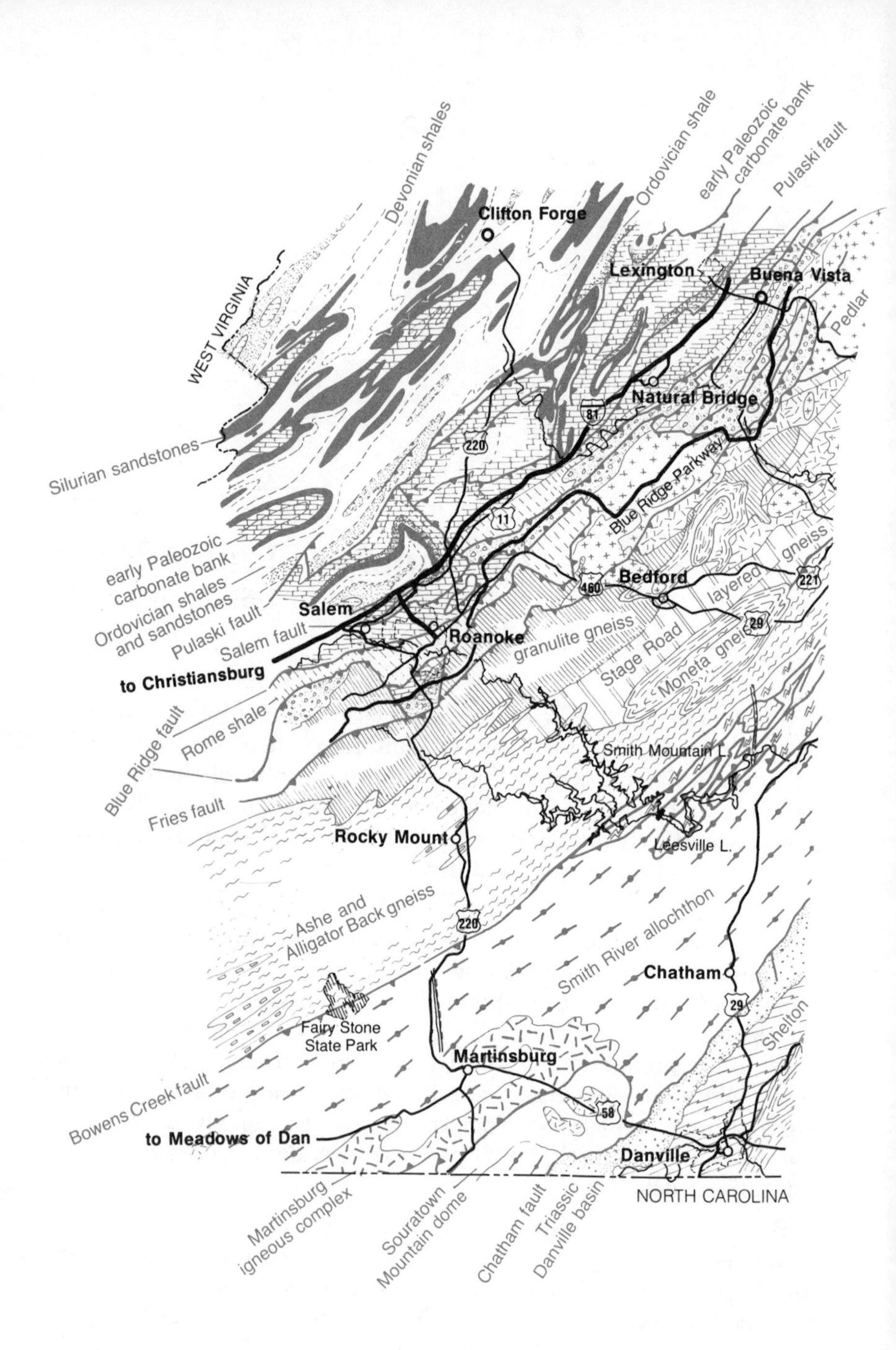

Devonian shales
Ordovician shale
early Paleozoic carbonate bank
Pulaski fault
Clifton Forge
Lexington
Buena Vista
Pedlar
WEST VIRGINIA
Natural Bridge
81
220
Silurian sandstones
Blue Ridge Parkway
11
early Paleozoic carbonate bank
Ordovician shales and sandstones
Pulaski fault
Salem fault
Salem
Roanoke
to Christiansburg
460
Bedford
layered gneiss
221
29
granulite gneiss
Stage Road
Moneta gneiss
Blue Ridge fault
Rome shale
Smith Mountain L.
Fries fault
Rocky Mount
Leesville L.
Ashe and Alligator Back gneiss
220
Smith River allochthon
Chatham
29
Fairy Stone State Park
Shelton
Martinsburg
Bowens Creek fault
58
to Meadows of Dan
Danville
NORTH CAROLINA
Martinsburg igneous complex
Souratown Mountain dome
Chatham fault
Triassic Danville basin

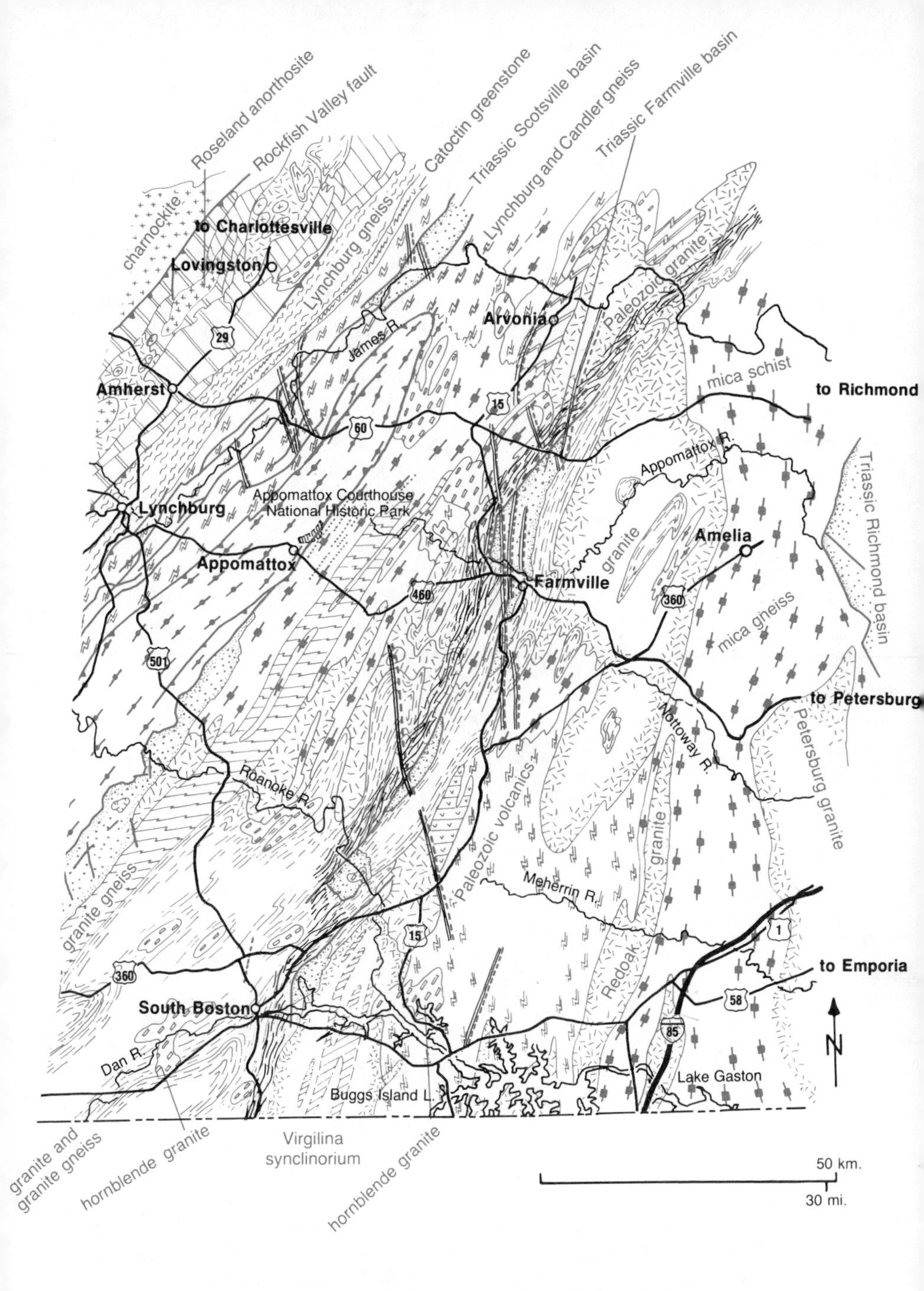

GREATER SOUTHSIDE PIEDMONT

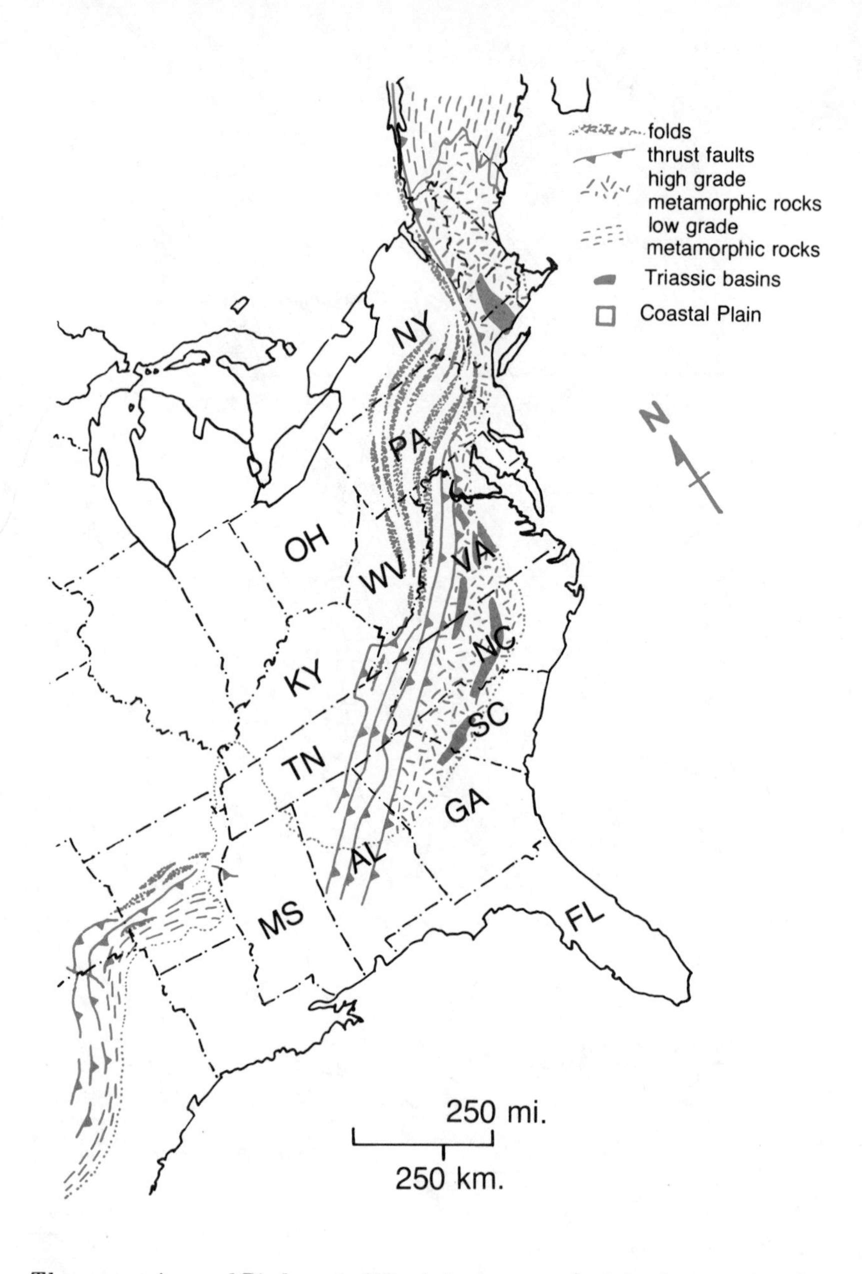

The mountains and Piedmont of Virginia are part of a belt of mountains that stretch from the Canadian Maritime Provinces on the northeast to the Coastal Plain onlap (dotted line) in Alabama. The relationship between the Appalachians and the Ouachita belt in Arkansas and Texas is obscured by the Coastal Plain sediments in Alabama and Mississippi. Major anticlines are stippled; major thrust faults have teeth on their upper edges; Piedmont is hatched for metamorphic rock and solid for Triassic basins.

VI
GREATER SOUTHSIDE PIEDMONT

HOW DEEP TO BEDROCK?

The bedrock of the Piedmont province, buried as it is by some 100 or more feet of saprolite, is hard for geologists or anyone else to see. And in summer, trees and shrubs hide all but the most spectacular natural outcrops. Moreover, these rare outcrops represent the most unusual rock types you can find in the Piedmont and are of relatively little help in deciphering the general geology of the region.

Old railroad cuts help a little, but railroad builders tried to avoid making deep, expensive roadcuts when they laid out their routes. Sparse outcrops are in the rivers and their tributaries, but virtually none on the high ground in between. Modern highways, quarries, and core drilling programs have helped geologists see the geology of the Piedmont, as have geophysical surveys which show the distribution at depth of dense rocks, magnetic rocks, and rock layers.

Mostly what you see of the subsurface when driving across the Piedmont is soil in freshly excavated saprolite. Deep red clayey saprolite suggests to you that the bedrock was a basalt, diabase, or amphibolite, rocks rich in iron and low in quartz.

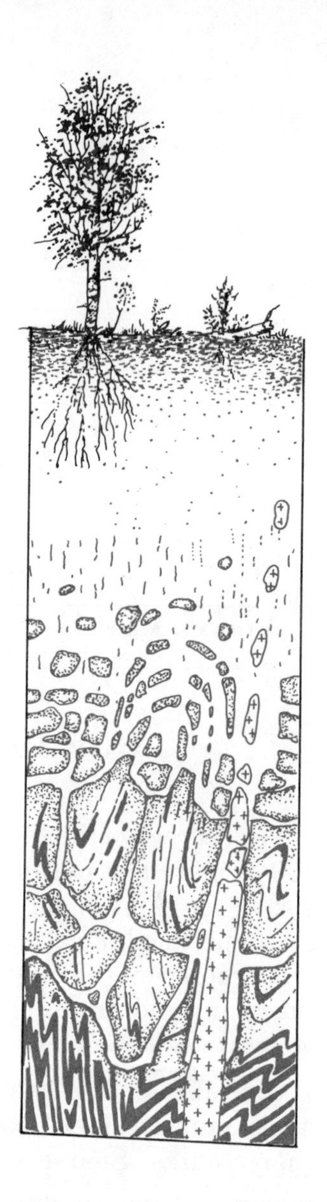

Rocks of the Piedmont geologic province are mostly buried beneath 100 feet or more of saprolite, a mantle of chemically weathered rock product grading downward through partially weathered rock to unaltered bedrock. Geologists trying to figure out Piedmont geology have to rely on sparse outcrop in river valleys and on interpretation of the saprolite.

Pale sandy saprolite suggests granite or some related rock that is rich in quartz and poor in iron-bearing minerals. Abundant mica flashing in the sun suggests that some kind of schist was parent to the present saprolite.

In spite of these difficulties, geologists have worked out most of the big pictures of the histories of the rocks of the Piedmont although many essential details still elude them. The oldest reliable radiometric ages for Piedmont rocks tell of igneous intrusion and intense metamorphism in Grenville times, about one billion years ago. Host rocks, intruded and metamorphosed by the granites and anorthosites are, of course, older; geologists estimate ages for them back to one and three-quarters billion years.

In addition to the core of the Blue Ridge anticlinorium, two areas of the Piedmont give Grenville ages—a granite intruding the State Farm gneiss just west of Richmond and the core of the Sauratown Mountain anticlinorium south of Martinsville. These are parts of an ancient continent upon or around which all subsequent geology developed.

From Grenville to Eocambrian, latest Precambrian, time, geologists can find very little record in the rocks to tell them what happened during that nearly half a billion years. But when rock deposition resumed in Virginia in Eocambrian time, a vast amount of cover—more than 20 miles—had eroded away to expose the once deep-seated Grenville plutonic and metamorphic rocks to Eocambrian oceans and atmosphere. And the Grenville topography was as deeply carved by streams and rivers as is the present Blue Ridge; geologists estimate 2000 feet of relief.

The Grenville continent rifted apart somewhere just east of the present Blue Ridge. Near this ancient shoreline, sands and gravels lagged behind in the valleys, today's Swift Run formation. Just offshore, a small rift, much like the Triassic basins of the Piedmont, filled wth muds and sands and gravels, today's Mechums River formation.

Layer upon layer of silt, sand, and gravel washed off the continent and came to rest on the newly-formed continental slope, today's Lynchburg and Ashe groups of formations. Farther offshore in the Iapetus Ocean, layers of mud came to rest on the ocean floor, today's Candler and Alligator Back formations. Toward the end of Lynchburg time, basaltic lava and volcanic ash began to spew out on the continent and its margins.

Catoctin basalts covered the Lynchburg, Mechums River, and Swift Run sediments and piled up against the granite mountains that stood above the sediments. Out in the Iapetus Ocean, sediments similar to the Lynchburg group appear to have surrounded or blanketed the Grenville Sauratown Moun-

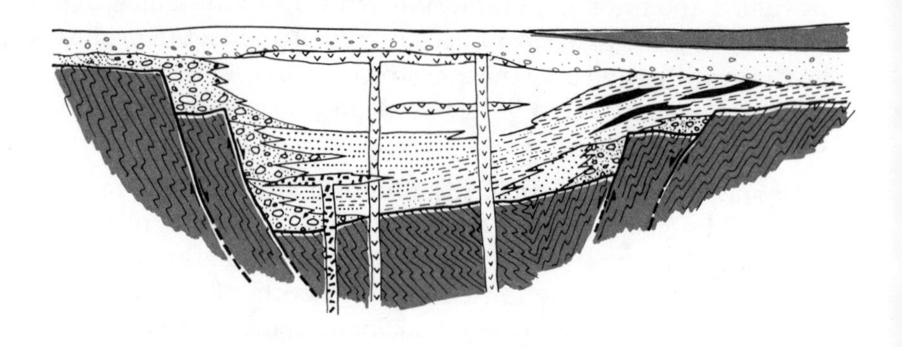

Rift basins formed in the Piedmont and in bedrock under the Coastal Plain during the Triassic and Jurassic just prior to the opening of the Atlantic Ocean. These basins are fault bounded, mostly with greater displacement on their western sides. They filled with lake and river sediments as the faults grew in size causing older sediments to dip more than younger ones. All the basins have basalt dikes and the Culpeper basin also has basalt flows. Coal has been mined from the Richmond and Danville basins and some of them may contain small oil fields.

tains. Similarly, basalts and sediments, now the metamorphosed Sabot and Maidens formations, blanketed the State Farm gneiss.

An alternative interpretation of the metamorphic rocks around the State Farm gneiss has them the host for Grenville intrusions, but altered beyond recognition by subsequent metamorphism. With this second interpretation, you have a separate terrane, called the Goochland terrane, stretching from Fredericksburg to the North Carolina state line east of the Spotsylvania lineament. In the Fredericksburg area this unit is called the Po River metamorphic suite and in North Carolina the Raleigh belt.

Still in the Iapetus, or ancestral Atlantic, Ocean near the beginning of the Paleozoic era, but west of the Spotsylvania lineament, layers of sediment interfinger with volcanic deposits, now the Chopawamsic formation. The eastern extension of this formation is known as the Ta River metamorphic suite in the Fredericksburg and Lake Anna areas. Along the James River, the equivalent to the Ta River suite is known as the Hatcher complex, which appears to be a volcanic pile intruded by its own magma source, the Columbia granite and the Ellisville granodiorite.

The rock formations that constitute the Smith River allochthon were being deposited at about this time—Eocambrian to Cambrian—as sediments in the Iapetus Ocean, somewhere southeast of the Sauratown Mountains. The types, structures, and metamorphic history of the rocks in the Smith River allochthon correlate well with North Carolina's Inner Piedmont belt of which it appears to be a transported slice.

In the area south of the nation's capital, at least two allochthons, the Piney Branch and the Potomac River, along with a slice of oceanic rock, the Sykesville ophiolite and melange, had been thrust upon the present day schists and sandstones of the Annandale group. Now folded into this are the Popes Head phyllites and metasiltstones. In Cambrian time, the Occoquan granite and related plutons intruded these rocks.

As sandstones of the Chilhowee group gave way to the early Paleozoic carbonate bank west of the present Blue Ridge, deposition continued in parts of the Iapetus Ocean. On top of the Chopawamsic formation in the Virgilina synclinorium, volcanic rock interbedded with clays became schists and slates of the Hyco and Aaron formations. These formations are the Virginia extension of the Carolina slate belt of North Carolina.

After some folding of the underlying Chopawamsic formation, sands followed by clays filled a series of synclines stretching from Quantico to Willis Mountain and beyond. Geologists consider the Quantico and Arvonia formations to be equivalent, now metamorphosed to slates, schists, and quartzites. Sparse fossils from these formations are of Ordovician age. These synclines produce slate around Arvonia, kyanite at Willis Mountain.

About 330 million years ago, the Petersburg granite intruded the eastern margin of the Piedmont and the basement rock underlying the adjacent Coastal Plain province. West of Richmond, the Fine Creek Mills granite intruded the State Farm dome at the same time. Most of the rock quarries in the Richmond and Petersburg area were blasted into these Mississippian granites.

To a geologist, metamorphism and magmatic invasion implies orogeny—mountain building. Granites, gneisses, and ore deposits, as well as lofty peaks, are involved in orogenesis. Lofty peaks may erode away, but ores and igneous and metamorphic rocks remain to reveal the mountain-building event. The lofty peaks are gone from the eastern Piedmont geologic province, but the granites and gneisses remain as their witness.

Rocks of the Virginia Piedmont reveal five separate orogenies. First, there is the billion-year-old Grenville event. Some geologists see evidence for an early supercontinent being assembled much as most of the Earth's continental rock was assembled into Pangaea at the end of the Paleozoic era some 280 million years ago. Certainly in the crystalline core of the State Farm gneiss, the Sauratown Mountains, and the Blue Ridge anticlinorium there was an overthickening and overheating of the crust, probably due to continental collision.

Granitic magma again invaded the core of the Blue Ridge anticlinorium in Eocambrian time. These magmas again represent another overheating, this time perhaps related to heat from rising basaltic magma that was to break through the crust as Catoctin volcanic rock. The opening of the Iapetus Ocean was an event of tension, not of compression, so such granites are called anorogenic. Apparently concurrent events in the Fredericksburg complex, however, were compressional, indicating that it may have been somewhere other than North America at that time. An additional difference in the Fredericksburg complex is that the Occoquan granite intruded it in Cambrian time.

As the Iapetus Ocean widened to its maximum in early Ordovician time, the early Paleozoic carbonate bank grew on its northern, tropical margin; North America faced south near the equator at this time. The African and European shore was 2000 miles away to the south, too far to be contributing sediments to the vicinity of North America.

But look at our present oceans. They are littered with microcontinents, volcanic islands, submarine volcanoes, plateaus, and banks, many which shed sediments locally. Various parts of the present Piedmont may have been just such geologic features of the Iapetus Ocean basin. And to close the Iapetus

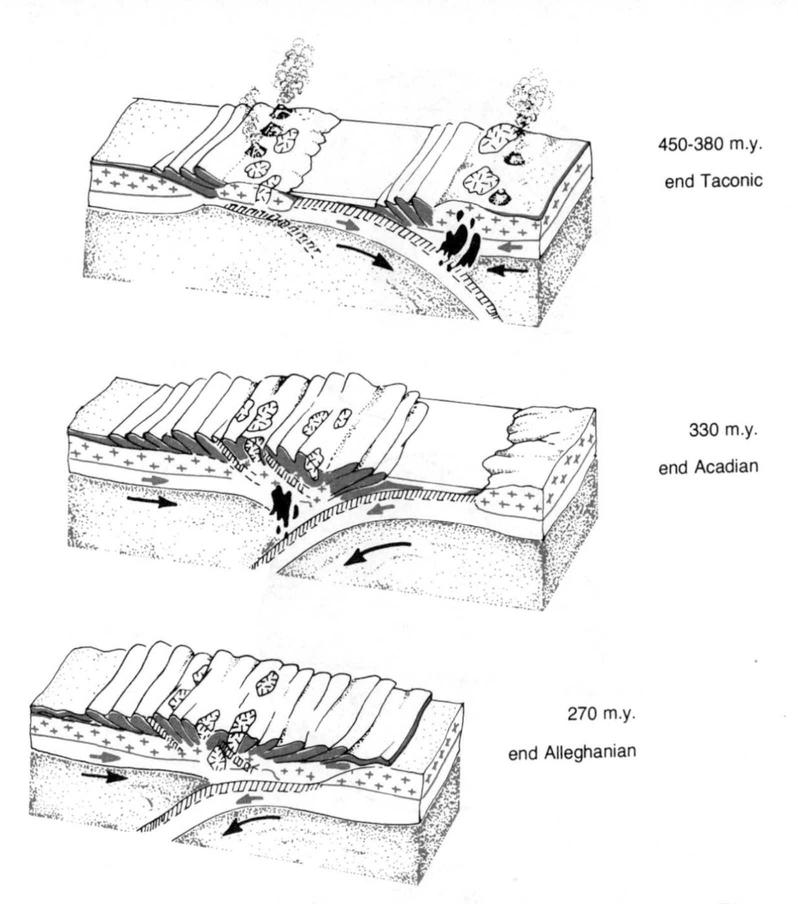

Three orogenies punctuated the Paleozoic in eastern North America. The Taconic orogeny, shown at top, added part of the Piedmont geologic province and folded it. The Acadian orogeny, middle, added more Piedmont, folded it again, and was accompanied by volcanic eruptions and plutonic intrusions. The Alleghanian orogeny, bottom, closed the Iapetus Ocean, thrust the inner Piedmont westward, and folded the Valley and Ridge geologic provinces.

Ocean, subduction would be attended by a volcanic island arc much like our Aleutian Chain or a trench-bounded continent much like western South America with her Andes Mountains.

In middle Ordovician time, the Taconic Ranges were shoved onto the North American continent in what is now eastern New York. Perhaps some part of the Piedmont bumped the edge of the North American continent and shoved the Smith River allochthon over the Sauratown Mountains and onto the east flank of the Blue Ridge. We do know that, during the Taconic orogeny, the early Paleozoic carbonate bank in the Valley and Ridge geologic province was snuffed out by drowning due to

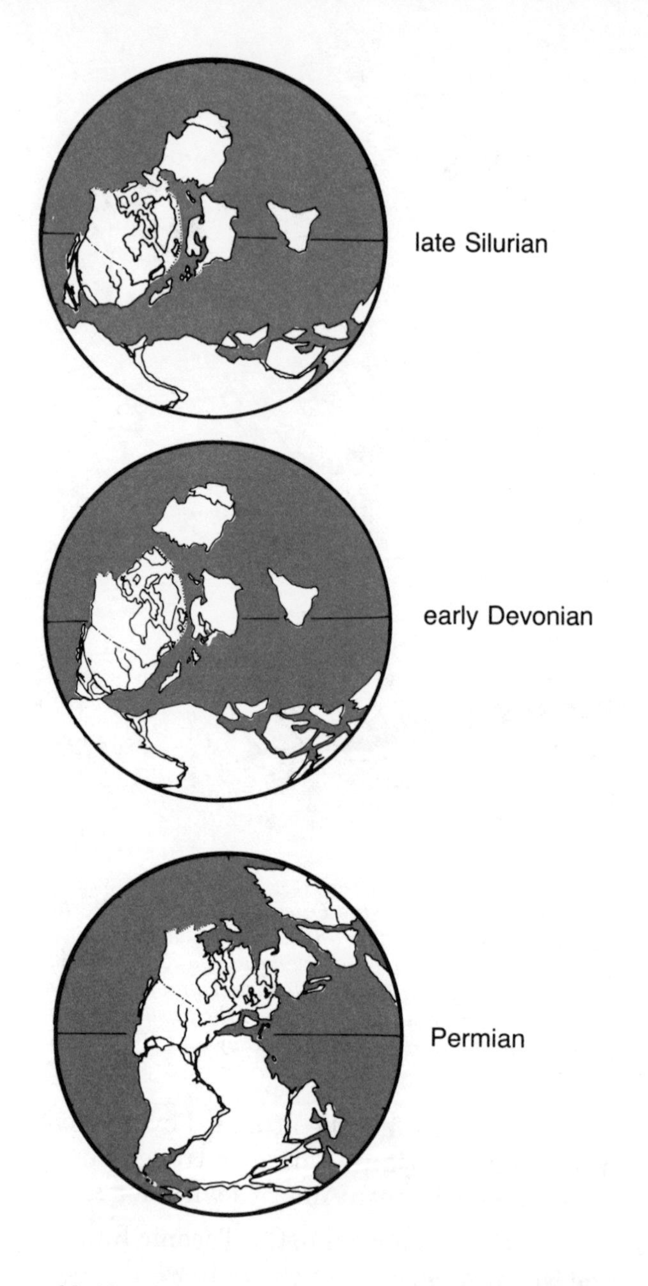

Pangaea assembled by plate tectonics at the end of the Paleozoic as shown in the lower map for the Permian. Going back to early Devonian on the middle map you can see part of eastern North America is still in the Iapetus Ocean and Europe is not attached to Asia. The top map shows the arrangement of the continents and microplates in the late Silurian. Note that eastern North America was near the equator throughout most of the Paleozoic.

depression of the continental margin; sediments poured into this new trough from what had been offshore.

By late Silurian time, that offshore source of sandy sediment had disappeared and a second, the middle Paleozoic, carbonate bank had established itself in the Valley and Ridge. But toward the end of Devonian time, a second crustal fragment docked on the North American continent with its greatest impact on Maritime Canada. The result of the Acadian orogeny in the Virginia Valley and Ridge was the drowning of the carbonate bank and another surge of sediments eroded from newly elevated land that had been offshore to the east. Perhaps something now hidden beneath Coastal Plain sediments and their offshore extension had docked on North America.

This Devonian trough, too, filled with sediments as the offshore source eroded. A third carbonate bank became established over some of the area in early Mississippian time. This was succeeded by the coal measures and their attendant marine and non-marine sediments recording cyclic rise and fall of sea level. Whatever caused the Mississippian plutonic activity in the eastern Piedmont seems to have had no effect on geologic processes to the west. The Spotsylvania lineament and the Hylas zone may be faults initiated just after intrusions. Perhaps part of the Piedmont was still not docked on North America, but was either in the Iapetus Ocean or attached to Africa on the far side.

Then came the final closure of the Iapetus Ocean in Permian

Beginning in the late Triassic, Pangaea began to split apart and the modern Atlantic Ocean basin began to open. The diagram shows schematically the symmetrical array of folds in North America and Africa before erosion and deposition of Coastal Plain sediments.

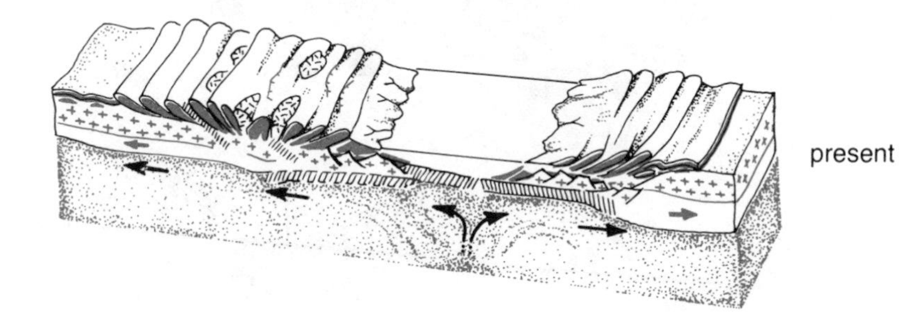

time, about 280 million years ago, with the docking of Europe and Africa onto the North American continent—the final assembly of Pangaea. The result was folding and thrusting of formations from the crystalline basement formations under the Coastal Plain to the Appalachian Plateaus and beyond. Not only was one formation thrust upon another, but continued compression caused initially near-horizontal fault planes to be turned over past vertical to the northwest. The Alleghanian orogeny put the finishing touches on most of the geologic structures you see today.

But a single world continent is too hot to stick together for long. After about 50 million years, continental rifting began in Triassic time in the Piedmont and the crystalline rock beneath the present Coastal Plain. Rift basins filled with sediments

Extension brought about by the opening of the Atlantic Ocean at the end of the Triassic caused graben formation. Most Triassic Piedmont basins are half grabens. One side, usually the western, is dropped considerably more than the eastern side.

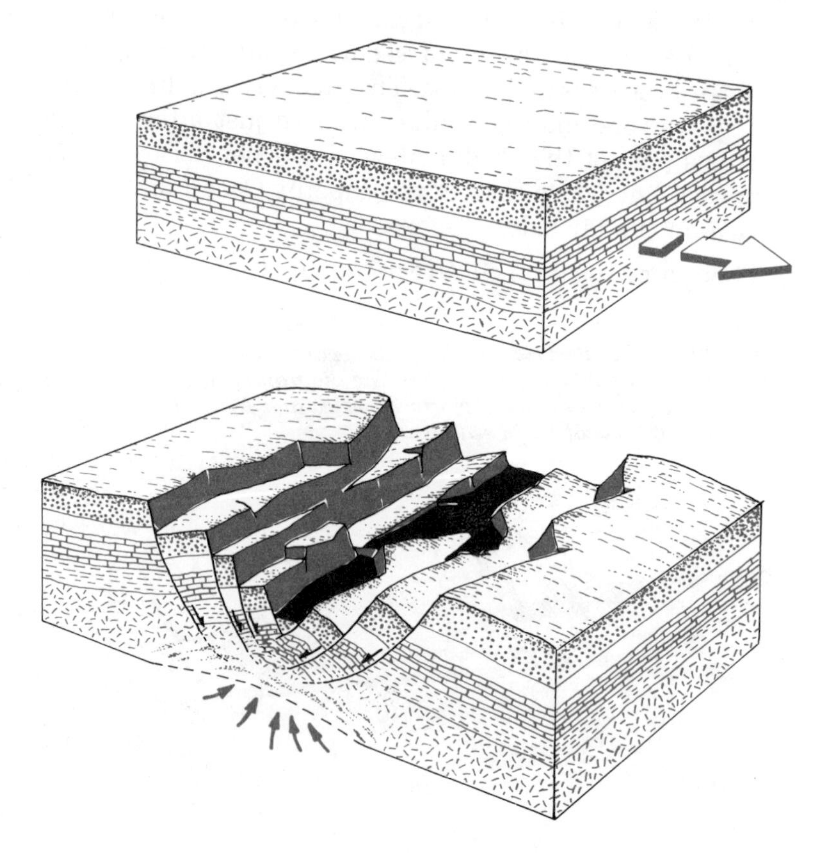

almost as fast as their bottoms dropped, except for the one that was destined to become the Atlantic Ocean. In Jurassic time, basaltic magma rose in all of them, but only continued in the Atlantic Ocean where it continues to do so today at the mid-ocean ridge.

The new Atlantic Ocean with its thin new basaltic crust required a transition zone between it and the thicker continental crust of North America. As the Atlantic rift formed, continental crust necked down by faulting along the older Spotsylvania lineament and Hylas zone to meet the new ocean crust giving the newly formed continental margin an eastward drainage. Erosion carved the Piedmont topography you can see today.

But how many once separate out-of-place masses of terrane make up the modern Piedmont geologic province? The geological jury is still out on that question.

WHAT IS THE TIME BY THE ROCK?

Long before geochronologists developed a radiometric time scale for the history of the Earth, geologists had established a scale of relative time and had named divisions of it. They could say that Cambrian time came before Pennsylvanian time just as anyone can say that Elizabethan time came before Victorian. Rocks from widely separate places are correlated by the fossils in them that grew at that time.

In a stack of sedimentary rock layers—strata—the layer at the bottom is older than all overlying strata. A dike or volcanic plug cutting through such a stack is younger than the stack. By making such observations world wide, geologists determined the order in which geologic events took place long before geochronologists were born.

Names for divisions of geologic time come from the places where the record in the rocks was first deciphered. Hence, rock formations from Devonshire in England give the name Devonian to all rocks formed during the same time and to that period of time. Pennsylvanian time is from the state of that name. Geologists use time names and radiometric ages for rocks as is convenient.

The Earth formed as a planet approximately 4700 million years ago. The oldest rock found on Earth crops out in southwestern Greenland; it is 4000 million years old. No rocks of this antiquity crop out in Virginia; her oldest rock is only about 1700 million years old.

Where do these numbers come from and what do they mean? Geochronology is the science of determining when a rock formed by measuring its radioactivity, either that of the whole rock or of a single mineral. Uranium and thorium are useful radioactive elements if present in sufficient quantity; geochronologists also use radioactive potassium, rubidium, and samarium.

Radioactive decay proceeds at a fixed rate much like interest on your money in a savings bank. The more capital you put in the account, the more interest you get. The more radioactive element in a rock or mineral, the more atoms decay per year. But while both capital and interest are in dollars, radioactive decay produces a new daughter element; uranium decays to lead and potassium to argon. And interest is positive, decay negative, decreasing the amount of the parent element.

Identical mathematical equations tell either story. If you know the amount of original cash deposit, the fixed rate of interest, and how many years the deposit will be left in the account, you or your friendly banker can calculate the amount you will withdraw at maturity. Geochronologists know the amount of the original deposit from measurements of other isotopes. They measure the amount there now, at the time of

The Petersburg granite has a radiometric age of 330 million years, a Mississippian age.

withdrawal. And they know the decay constant, negative interest rate. Hence, they can calculate the time elapsed since the original deposit was made. This time is a radiometric age.

A geochronologist measures the amount of radioactive element, capital, and the amount of daughter element, interest, in a rock or mineral. Unlike interest rates, which seem to be always changing, decay constants do not fluctuate and have been measured very accurately. Given the decay constant, interest rate, and the amount of parent and daughter, a geochronologist can calculate when a rock or mineral formed and give that number as the radiometric age of the rock, sometimes inaccurately called the absolute age. Thus, many rocks contain built-in chronometers.

Rocks that make up a large part of the Blue Ridge province, the State Farm dome, and the Sauratown Mountains formed in Grenville time, about 1200 million years ago. They intrude a host rock 1700 million years old, the oldest yet found in Virginia.

EVERY WHICH WAY IS NORTH

A compass needle aligns itself with the Earth's magnetic field and points toward a magnetic pole, north or south. Although the magnetic poles are not today exactly aligned with the Earth's axis of spin—the geographic poles—the magnetic axis does coincide with the spin axis when averaged over thousands of years.

Magnetic minerals, chiefly magnetite, "remember" how the Earth's magnetic field lay when and where they formed. Igneous and some metamorphic rocks remember the Earth's magnetic field as it was when the mineral cooled through its magnetic inversion temperature. Magnetic sediment grains align themselves with the Earth's magnetic field as they settle to the bottom. Once fixed in a rock, this magnetism maintains its original alignment even when the magnetic poles move.

How strong is this rock magnetism? One geologist had the task of measuring the extents of two basalt lava flows in the state of Washington. Although he could tell them apart in the

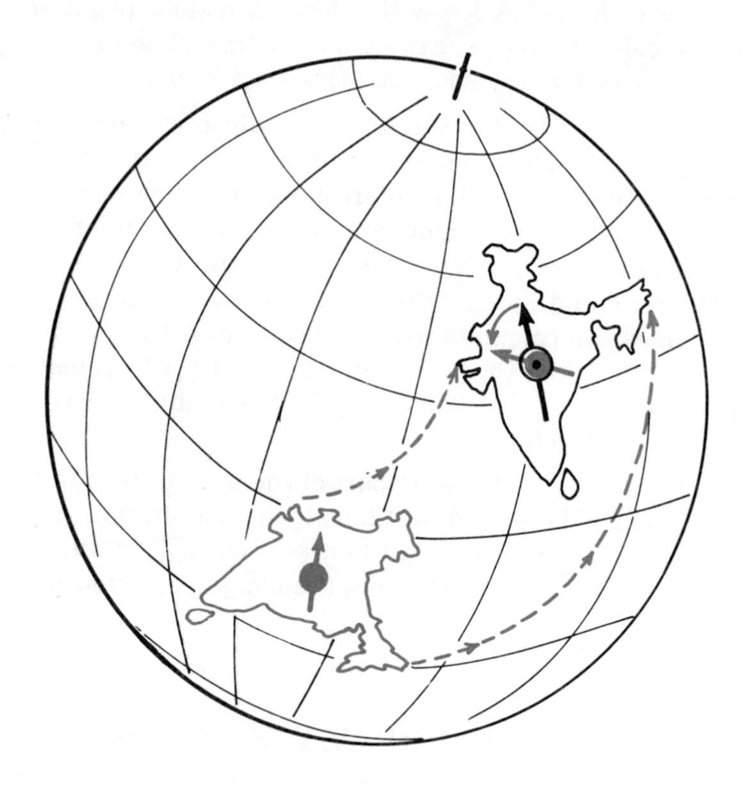

When magnetic minerals crystallize, they align with the geomagnetic field, as in the continent in color. When their crustal plate moves, their old magnetic alignment is preserved and is out of line with the current geomagnetic field, in black.

laboratory and when they occurred in the field one on top of the other, he could not tell them apart when only one was present at an outcrop. After weeks of frustration, he happened to set his compass on the older flow. To his astonished delight, the needle pointed south. When set on the younger flow, it pointed north. From then on he could tell the two flows apart just as fast as he could set his compass down.

Although most rock magnetism is considerably weaker than this example, magnetometers much more sensitive than a pocket compass exist. Recent studies of rock magnetism have led to several startling discoveries.

First, every few hundred thousand years the Earth's magnetic field weakens during a thousand years or so and then reverses itself. Calibration of these reversals with radiometric "clocks" permits analysis of geologic events or processes. The opening of the Atlantic Ocean, for example, is recorded in alternating paleomagnetic stripes on either side of the mid-Atlantic ridge. Even ancient archeological sites may be "dated" by using magnetic minerals in a fire pit.

Second, the geomagnetic field is tied to, and therefore controlled by, the Earth's spin when averaged over thousands of years. This gives geologists a fixed reference point for paleogeographic studies. In Cambrian times, for example, the equator ran roughly parallel to the Mississippi River with Virginia just to the south.

Third, plotting paleopoles through time shows them to migrate. This progression is called polar wander even though the poles do not wander; the entire crust and upper mantle—the lithosphere—of the Earth shifts with respect to the spin axis. This outer rocky portion is not attached to the lower mantle and core of the Earth!

Fourth, the paleopoles of different continents follow different tracks. Since a globe can have only one axis of spin, this divergence of poles indicates how far different lithospheric plates have moved from one another. Going backward through time toward the Jurassic age, the only way to keep the North American polar track coincident with its European counterpart is to close the Atlantic Ocean, a couple of inches every year, thereby reassembling the supercontinent of Pangaea.

WHAT IS SUSPICIOUS ABOUT SUSPECT TERRANES?

Geologists refer to parts of the bedrock of the Blue Ridge and Piedmont geologic provinces as suspect terranes. That is to say, they have reason to suspect that the rock formed elsewhere and was moved along a fault to its present location. In most cases they suspect that once an ocean basin lay between that area of rock and its neighbors and that subduction consumed the intervening oceanic crust.

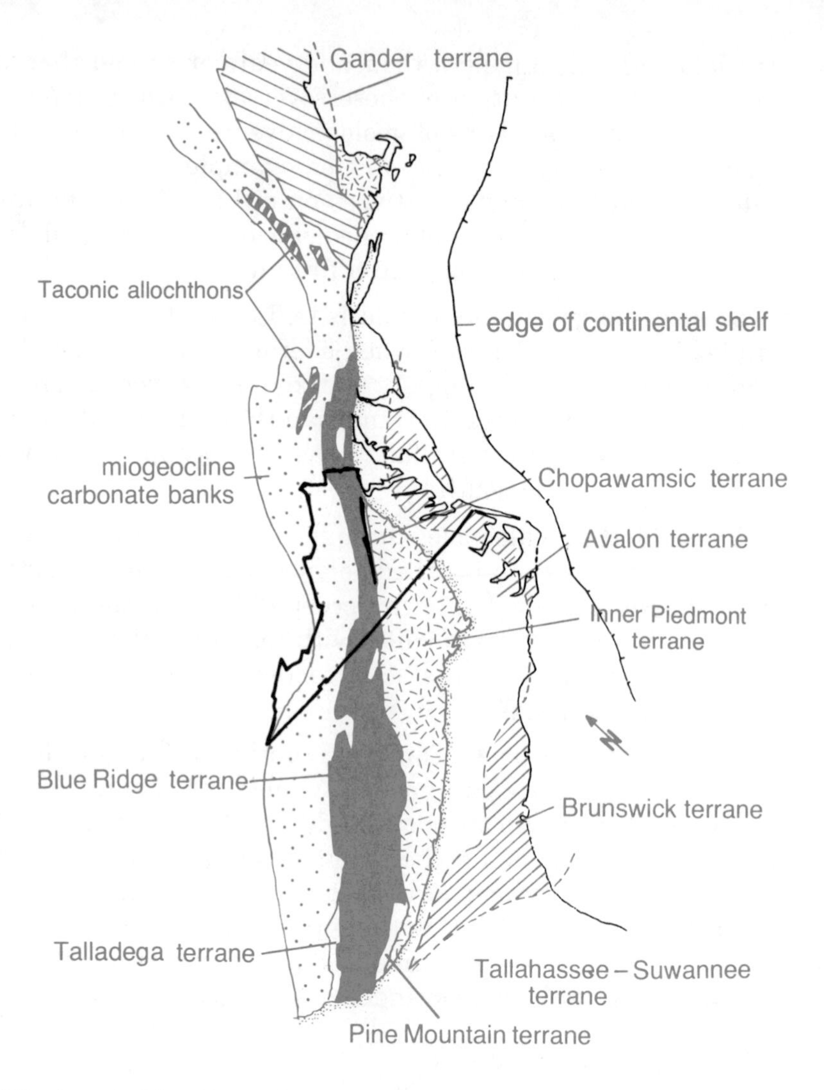

The Appalachian Mountains are made up of a variety of terranes, many of which had their early geologic histories somewhere other than on the North American continent. The miogeocline has been North American since its beginning. Terranes which are known to have originated elsewhere are allochthonous. Those which may have originated elsewhere are suspect.

In some cases, geologists suspect that horizontal fault movement transported the rock along the coast much like California west of the San Andreas fault is slipping along the West Coast. Or one slab of rock may have been thrust into place over other rocks with the thrusting rock now eroded away.

Two major observations lead geologists to suspect a terrane. First, faults bound it all around and geologic features cannot be traced to neighboring terranes. Second, some of its geologic history, especially its early history, is completely different from its neighboring terranes. Geologists look especially for differing metamorphic histories.

Suspicions are confirmed when study of its paleomagnetism reveals that a terrane appears to have magnetic poles that were different from those of its neighbors. When suspicions are confirmed it is labeled allochthonous. The Chopawamsic terrane is suspect. The Smith River allochthon is no longer considered a suspect terrane. Geologists are sure that it was transported into place from another place of origin.

WHENCE THE SMITH RIVER ALLOCHTHON?

The Smith River allochthon lies between the Ridgeway and Chatham faults on the east and the Bowens Creek fault on the west. These faults converge north of Appomattox. The boundary faults also converge to the south in North Carolina where they disappear into the Brevard zone. This slice of rock is more than 150 miles long and as much as 20 miles wide.

Since the eastern and western faults merge at depth, the allochthon is rootless in that it is detached from rocks which originally underlay it. Three episodes of metamorphism can be detected in the metasedimentary formations of the Smith River allochthon; the adjacent post-Grenville sediments in the Sauratown Mountains and Blue Ridge are not so deformed.

Two sequences of rock make up the Smith River allochthon, one a sedimentary sequence with some volcanic rocks in it, the other intrusive igneous rocks. The layered sequence is probably Eocambrian to Cambrian in age or 600 to 500 million years old; the intrusive suite is Ordovician or around 450 million years old. The allochthon was thrust to its present location west of the Sauratown Mountains after the igneous rocks had crystallized.

The oldest rock formation in the allochthon is the Bassett formation which is made up of mica gneiss and amphibolite.

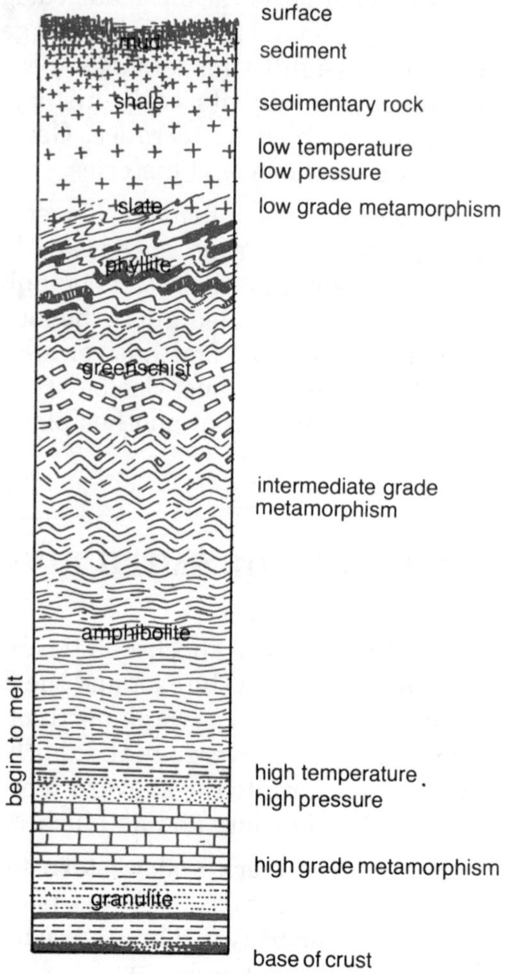

surface
sediment
sedimentary rock
low temperature
low pressure
low grade metamorphism
intermediate grade metamorphism
high temperature
high pressure
high grade metamorphism
base of crust
mud
shale
slate
phyllite
greenschist
amphibolite
granulite
begin to melt

The kind of metamorphic rock that you see depends first upon its initial composition and second upon the pressure and temperature of recrystallization—its metamorphic facies. The grade of metamorphic facies increases from top to bottom.

The mica gneiss probably represents metamorphosed sedimentary rocks and the amphibolite is metamorphosed basalt flows and ash falls. Numerous small, now altered, mafic intrusions invade the Bassett formation.

Overlying the Bassett formation is the Fork Mountain formation with two different rock types—garnet-mica schist and garnet-mica gneiss. Both units were again metamorphosed under less intense conditions of pressure and temperature.

Igneous intrusions of the Martinsville igneous complex invaded large areas that are now in the western part of the Smith

River allochthon. The Rich Acres formation, predominantly gabbro, and the Leatherwood granite cross cut each other, which indicates to geologists that they are contemporaneous. Simultaneous intrusion of gabbro from the mantle and granite from the crust of the Earth is not unusual.

Whence the rocks of the Smith River allochthon? They most resemble rock formations of the Inner Piedmont belt of the Carolinas and so may have come from there. But some geologists think the Inner Piedmont is also allochthonous. Thus, they come from wherever the Inner Piedmont came from.

When were the rocks of the Smith River allochthon slid into place? Age dates indicate that it was emplaced during the Taconic orogeny about 450 million years ago.

IS THE CHOPAWAMSIC TERRANE SUSPECT?

The Chopawamsic terrane extends as a narrow belt of rock formations for about 100 miles between Fredericksburg and the vicinity of Farmville. It contains the Quantico, Columbia, and Arvonia synclinoria. The eastern boundary of the terrane is the Spotsylvania lineament and the western is the Vandola and Chatham faults.

The Chopawamsic formation consists of metamorphosed light- and dark-colored volcanic flows and ash falls with lesser amounts of sandstone and phyllite to a thickness possibly as great as 10,000 feet. After deposition, the formation was folded slightly into a trough which then filled with more sediments. Metamorphism converted these sediments into the slates of the Quantico formation, slates and quartzites of the Arvonia formation, and the kyanite quartzites that form Willis Mountain.

One reason to suspect that the Chopawamsic terrane was transported is that the few Ordovician fossils found in the Arvonia slates do not resemble those from the lower Paleozoic carbonate bank to the west. They more closely resemble European fossils. Another reason is the presence of an ophiolite melange west of the north end of the terrane. A third reason is that its eastern boundary, the Spotsylvania lineament, may represent a major Paleozoic fault or suture. Finally, much of

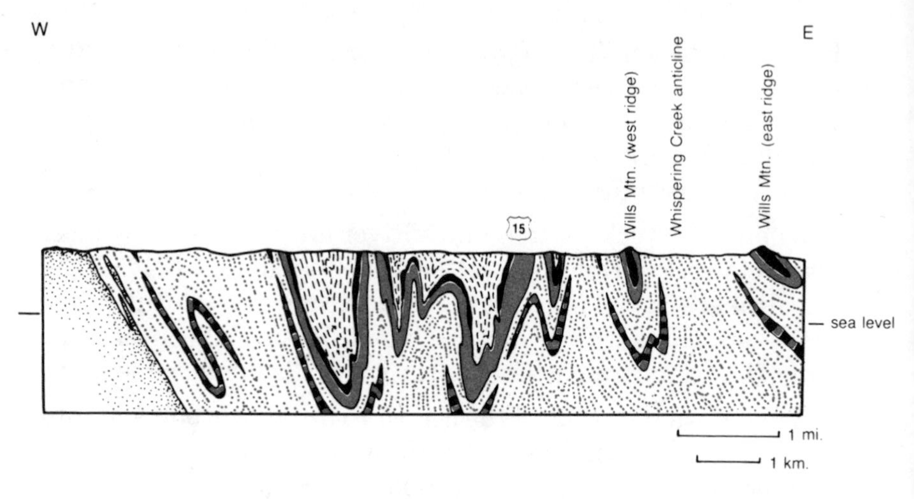

The cross section of the Arvonia synclinorium at Willis Mountain shows synclines of Ordovician infolds into Cambrian Candler formation.

the western boundary of the Chopawamsic terrane is a fault.

One reason to suspect that the Chopawamsic terrane was not transported significantly is its apparent transitional sedimentary lower boundary with the underlying Candler formation in central Virginia. Resolution of these suspicions requires further geologic research but the terrane is suspect.

WHAT IS THE SPOTSYLVANIA LINEAMENT?

Since you cannot see the Spotsylvania lineament in very many places except as a line on a map, you might well wonder, who cares? But that lineament, whatever it may represent, runs close to the North Anna nuclear power plants. And we, the People, like to know the stability of the ground around our nuclear power stations. Consider three points:

First, the Spotsylvania lineament is not entirely invisible to the careful observer. To the south, it is the western border fault of the Triassic to Jurassic Danville and South Boston basins. Hence, land southeast of it moved down as the Atlantic Ocean basin opened.

Next, on its extension underneath the sediments of the

Coastal Plain geologic province, these Tertiary sediments are shoved up and west against the Piedmont rocks. Hence, the Spotsylvania lineament appears to have controlled the location of minor faulting about 50 million years ago.

Finally, on aeromagnetic maps that show variations in the intensity of rock magnetism, the Spotsylvania lineament separates rocks with entirely different patterns of rock magnetism.

What does all this mean? Nothing appears to have happened along the Spotsylvania lineament for several tens of millions of years. Parts of it guided faulting 50 million and 200 million years ago. And the Spotsylvania lineament just may be an important fault or suture between Paleozoic microplates.

WHY ARE PIEDMONT MOUNTAINS?

Although the Piedmont geologic province of Virginia has a rolling landscape with deeply weathered bedrock, it has a few prominent mountains, pieces of real estate that stand well above the surrounding terrain. What weather-resistant materials protect these sentinels from being washed away?

Catoctin, Bull Run, and Southwest mountains are underlain by greenstone, a rock that began as molten magma that erupted onto the land. Textbooks tell you that dark-colored igneous rock, such as basalt, weathers faster than light-colored igneous rock, such as granite.

But in the western Piedmont, the folding, faulting, and metamorphism that ended the Paleozoic era altered the rocks. The biggest effect on granites and granite gneisses was to crush the grains and leave cracks that ground water could get into. The biggest effect on the basalts and their ilk was a thorough recrystallization that made them mechanically impervious rocks called greenstones. Hence, we have greenstone mountains. Try collecting a sample of Catoctin greenstone from a roadcut with a hammer!

Willis Mountain is held up by weather-proof quartzite but human erosion of it for its kyanite content far exceeds nonhuman erosion rates. This beacon north of Farmville will, within the lifetime of many readers, succumb to our need for sparkplugs.

Other Piedmont mountains are underlain by mica-rich and feldspar-poor rock formations such as schists and phyllites. The late Precambrian Candler phyllites hold up Smith Mountain as well as Candler Mountain. Other hill-forming rocks include pyroxenites and olivine-rich dunites and their alteration products. Rocks rich in feldspar tend to underlie topographic low areas.

One exception to this rule about little feldspar in Piedmont mountains is White Oak Mountain which has many beds of arkose, a feldspar-rich sandstone. But since these formations are late Triassic, some 210 million years old and 100 million years younger than the Alleghanian orogeny, those feldspar grains have not been crushed.

Triassic sandstone, arkose, and conglomerate hold up White Oak Mountain north of Danville. Unlike Paleozoic rocks containing feldspar which was crushed during periods of mountain building, these are uncrushed and, therefore, weather resistant.

Petersburg granite pocked with potholes at Belle Isle Park in Richmond.

US 60
Richmond — Amherst
88 mi./142 km.

The route begins at the Fall Line where Coastal Plain sediments lap onto the crystalline rocks of the Piedmont. Whereas the Coastal Plain formations are rocks yet to be, the Piedmont formations are blanketed with rock that used to be. Chemical weathering of the rocks left a blanket of saprolite more than 100 feet thick over much of the province. The highway tends to follow the high ground between the James and Appomattox rivers—high ground where the saprolite is the deepest. About all you can see for most of this leg is changes in soil color in fresh excavations and plowed fields.

Midlothian is on the eastern edge of the Richmond basin, which opened with the Atlantic Ocean. It contains a fill of Triassic and Jurassic sedimentary rocks. The basin here is only about 7 miles wide. Most of Powhatan County is underlain by mica schists, eastern Cumberland County by granite. The change in bedrock does not affect the appearance of the landscape.

A few miles north of US 60, about midway between Midlothian and Powhatan, a granite pegmatite was worked briefly during World War II for its mica content. Fragments of beryl crystals up to five feet long were recovered then. More recently, topaz crystals of gem quality up to eleven inches long and weighing as much as nine pounds turned up at the same mine. Commercial interest, however, centers on the green amazonite variety of feldspar and on tantalum minerals.

Two miles east of the line between Cumberland and Buckingham counties, the route crosses the northern end of the Danville basin, another of those that formed as the Atlantic Ocean opened. To the south stands Willis Mountain, a ridge of resistant quartzite that contains kyanite. Very small outcrops related to this ridge appear along the road at Whispering Creek. The folds at Willis Mountain are part of a synclinorium that stretches from there to Fredericksburg.

High ground under the intersection of US 60 and US 15 is on the western limb of the Arvonia synclinorium. From here on a clear day you can see the Blue Ridge on the western skyline some 45 miles away.

Boulders and outcrops in the fields between Buckingham and the James River are from the Fork Mountain formation. The James River here mostly follows an outcrop belt of Lynchburg gneisses. Much of the Lynchburg formation is topographically low, possibly because the abundance of crushed feldspar in the gneisses makes them less resistant to weathering.

Willis Mountain towers above the Piedmont because of its weather-resistant kyanite quartzite bedrock. To the right, parts of the ridge have already been mined away. John Marr photo. Courtesy of the Virginia Division of Mineral Resources

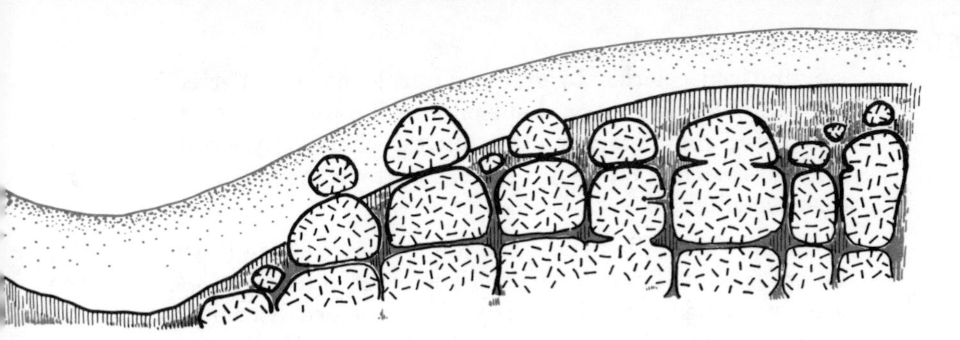

Rocks weather along fracture joints below the land surface. Later erosion in a nearby stream channel exhumes the unfractured, spheroidally weathered blocks of granite along its valley walls.

US 460
Petersburg — Roanoke
166 mi./183 km.

Between Petersburg and Lynchburg, US 460 crosses the full width of the Piedmont geologic province. Except near Lynchburg where the highway climbs out of the valley of the James River, bedrock lies beneath a thick blanket of saprolite and exposures are virtually nonexistent. The highway follows the "high" ground between the drainages of the Appomattox and Nottoway rivers. The saprolite, along with the dense vegetation, severely hampers geologists in their interpretation of Piedmont geology.

Petersburg stands where the Appomattox River crosses the Fall Line. The Fall Line is not straight, and the Appomattox River flows on granite where the bluffs are underlain by Tertiary sediments and terrace deposits. Quarries, such as the one where US 460 and Interstate 85 join west of Petersburg, simply remove the sediments to get at the granite. The Petersburg granite is a pink to gray rock composed of about two-thirds feldspar and one-third quartz, with a sprinkling of mica.

Between Petersburg and Burkeville, the topography has a gentle roll but much of it is level enough for row crops. The soil is pale tan in plowed fields and fresh excavations indicating a light-colored granite or granite gneiss at depth. Near Blackstone, the route crosses the Hylas fault zone if it extends that far south. The quarry at Burkeville is in granite.

Between Burkeville and Farmville, the topography is more deeply dissected than farther east and less land is cleared. Crystals of quartz ranging from colorless to amethyst purple may be collected from some private fields near Rice. Feldspars of the underlying granite have weathered completely to clay, but the chemically resistant quartz crystals turn up in the soil, especially after a shower following spring disk plowing.

On the Farmville bypass the route crosses a small Triassic basin. Before erosion it may have been connected at a higher level with the Farmville basin to the north. The western border fault of the Farmville Basin is in line with the Spotsylvania lineament, an important but not entirely understood feature in the Piedmont. Some geologists think the Spotsylvania lineament is a boundary associated with collision of the Avalon terrane to the east with the already docked Piedmont terrane to the west.

This boundary was reactivated in Triassic time when the Atlantic Ocean was just beginning to rift open, according to this line of speculation, and crustal rock dropped to form a trough as the new edge of the continent necked down to meet the new ocean basin. Sediments washed in, mostly from the northwest side during Triassic and Jurassic time. Because filling with sediments kept pace with faulting of the basin margin, the topographic basin was never very deep.

Willis Mountain stands as a sentinel above the surrounding landscape 10 miles north of Farmville. This southern extension of the Arvonia syncline resisted chemical erosion because of its mineral composition, chiefly kyanite and quartz. But because kyanite is raw material for the manufacture of porcelain, human erosion of Willis Mountain may change the skyline. You can glimpse this unusual peak through breaks in the trees between Farmville and Elam.

At Elam, deep, red saprolite in the railroad cuts on the south side of the highway should tell you why there are so few rock outcrops on the Piedmont. Whatever parent crystalline bedrock lies at depth is now chemically altered to clay and red hematite.

The Nottoway River cascades over Redoak granite at The Falls.

This railroad cut west of Farmville reveals deep red clay of decayed bedrock.

At Appomattox, the route crosses another major Piedmont boundary, the Chatham fault, which defines the eastern edge of the transported Smith River allochthon. This unit, up to 20 miles wide, extends from 25 miles north of the route southward into North Carolina where it ends in the Brevard zone, a distance of 160 miles. Formations in the allochthon are mostly schists and gneisses, which have been deformed and metamorphosed at least three times and then intruded by granites and gabbros of the Martinsburg igneous complex. They were then transported by faulting at least 30 miles over the Sauratown Mountains to their present location, possibly during the Taconic orogeny in late Ordovician time.

Near Concord, the route crosses the western border of the Smith River allochthon at the Bowens Creek thrust fault. Here, the allochthon was shoved onto gneisses and greenstones of the Lynchburg group. Only a couple of miles west, the Lynchburg is itself thrust upon the Candler formation, which holds up Candler Mountain to the south.

Lynchburg is on rocks of the Lynchburg group—schists, gneisses, amphibolites, sandstones, and a few marbles. The original sediments and lava flows were laid down on the North American continental margin in late Precambrian time, about 600 million years ago. They were folded, faulted, and metamorphosed to their present condition in Paleozoic time.

The route crosses onto the Moneta gneiss about two miles west of the Bedford County line. Geologists disagree as to whether the Moneta is a distinct part of the Lynchburg group or is a much older formation that has undergone extensive metamorphism. Between Lynchburg and Bedford, it appears to contain infolds of recognizable Lynchburg formations.

The route crosses onto layered gneisses of the Lovingston massif of the Blue Ridge province near Big Otter River. These gneisses were

metamorphosed and intruded by granitic plutons in the Grenville event approximately 1000 million years ago. Geologists use the term "event" when something has happened to the rock but so much of the geologic record is hidden or erased that they are not entirely certain of an explanation.

From west of Bedford you can see the Peaks of Otter rising in the north. In Colonial times they were thought to be the highest peaks in the state. West of Bedford, the route crosses the Rockfish Valley thrust fault, which separates the Lovingston granite on the east from the Pedlar granite on the west. Near Goose Creek, the route crosses the Blue Ridge thrust fault, which separates the Blue Ridge from the Valley and Ridge province.

The Blue Ridge thrust fault lies well east of the topographic Blue Ridge because here much of the Blue Ridge thrust sheet is eroded. Goose Creek cut through the Grenville granites and gneisses of the Blue Ridge thrust sheet and into the Cambrian Rome formation beneath. The surrounding hills are twice as old as the rock underfoot.

In the town of Blue Ridge you can look north into the Blue Ridge quarry in a folded and faulted thrust slice of limestone of the early Paleozoic carbonate bank. Exposures of limestone in this quarry give geologists a three-dimensional picture of rock conditions in a very small area and keep them wondering how much they miss elsewhere by seeing only scattered bedrock outcrops.

The interchange between US 460 and the Blue Ridge Parkway is on the Cambrian Elbrook carbonate of the early Paleozoic carbonate bank. From Coyner Springs to downtown Roanoke the route runs on Cambrian Rome siltstones and limestones. Roanoke lies on the eastern side of the Valley and Ridge geologic province.

This pedestal of Petersburg granite clearly resisted weathering and erosion. The black case is 3x5 inches.

US 58
Emporia — Meadows of Dan
179 mi./288 km.

Emporia is on the Meherrin River where it crosses the Fall Line. Here, sedimentary rocks of the Coastal Plain lap onto ancient granite of the Piedmont. Very young river gravels form terraces on both sides of the river. The route crosses those terrace deposits into Brunswick County, then passes onto terrain cut into Petersburg granite, which intruded the older rocks of the Piedmont in Mississippian time, about 330 million years ago. Small outcrops of granite appear on either side of Lawrenceville. Near the bridge over the Meherrin River, the road crosses onto the older mica gneiss that the Petersburg granite invaded. Here the landscape takes on the rolling appearance typical of large areas of the Piedmont province.

Just west of the intersection with US 1 north, the road crosses the northern tip of the eastern belt of Redoak granite. The intersection with US 1 south is in another belt of the same granite. A smaller belt of Redoak granite is just east of the bridge over the Roanoke River arm of Kerr Reservoir. Just east of Aarons Creek, US 58 is in more Redoak granite, here cut by a diabase dike on the north side of the highway. Just down the hill from this cut, the road crosses into the

Aaron formation in the Virgilina synclinorium.

The northern extension of the Carolina slate belt extends some 40 miles into Virginia as the Virgilina synclinorium. The Aaron slates, greenstones, and metamorphosed volcanic ash are the upper formation in the synclinorium. The lower formation contains up to 15,000 feet of light-colored volcanic rock, now slightly metamorphosed. Geologists think these rocks were deposited from a chain of volcanic islands during Eocambrian or Cambrian time.

Copper was mined in the greenstones of the Virgilina synclinorium from Colonial times until the onset of World War I. The copper minerals were mostly in quartz veins that cut the greenstones. Three-quarters of a million pounds of copper and more than one thousand ounces of gold came from the district, primarily south of US 58 and extending past Virgilina into North Carolina. Minerals may still be collected from some of the abandoned dumps on private land.

South Boston, the junction with US 360, is in a small Triassic to Jurassic basin, one of several in line with the large Farmville basin to the north. West of South Boston, the route crosses onto mica and hornblende gneisses which may be metamorphosed Chopawamsic formation. They underlie the Virgilina synclinorium.

Weather-resistant rocks cause riffles in the Dan River at Danville.

Danville is on the Shelton granites and granite gneisses, which crop out in the Dan River in Danville and in the roadcut at the intersection of US 58 and US 29 in Danville. The route continues across the Shelton rocks for three miles west of Danville before crossing the Vandola fault onto the Danville basin, here less than three miles wide.

West of the Danville basin lies the Smith River allochthon, a piece of the Inner Piedmont of North Carolina that was shoved over the Sauratown Mountains to its present location. For about one mile the route crosses formations of the allochthon before crossing the Ridgeway fault onto Grenville gneisses.

About 5 miles east of downtown Martinsburg, US 58 again crosses the Ridgeway fault onto the Smith River allochthon. This part was intruded by magmas of the Martinsburg igneous complex. West of Martinsville, the route crosses the much deformed sedimentary rocks of the allochthon. About a mile south of its junction with Virginia 8, the highway crosses the western boundary of the allochthon onto formations of the Blue Ridge province, here the pinstripe gneiss of the Alligator Back formation. US 58 climbs the Blue Ridge escarpment, here cut into gneisses and amphibolites of the Precambrian Alligator Back formation, to Meadows of Dan and the Blue Ridge Parkway.

US 501
Buena Vista — Lynchburg
26 mi./42 km.

Jagged Chilhowee foothills to the Blue Ridge overlook Buena Vista. Outcrops beside the road in the gorge of the James River expose the same Chilhowee sandstones that hold up the sharp ridges east of Buena Vista. Looking through the trees, one can see the same steeply dipping beds on the far side of the river. To the east, the cliff-forming beds arch over to horizontal in an anticline.

Chilhowee sandstones in the west flank of the anticline in the James River gorge through the Blue Ridge show minor folds.

Between Grenville granulite gneiss and the overlying base of the Chilhowee sandstones is a phyllite that geologists think may have been a Precambrian soil.

Roadcuts on the north side of the highway display many small folds that developed as the large anticlinal arch formed. One mile east of the county line where the road curves sharply at the head of a stream valley is a contact where Chilhowee sandstone lies on Pedlar granulite gneiss. Between the granulite and the sandstone is a zone of phyllite about 2 feet thick which appears to be metamorphosed soil formed on the granulite before the sandstone was deposited. This shows what can happen to ordinary dirt in 600 million years. About a mile and a quarter farther east, where the road comes out of the trees and onto the floodplain, a similar outcrop exists on the east side of the anticline. Folds in Chilhowee sandstone appear in the walls of the small quarries along the highway. The white sandstone just east of the quarries contains *skolithus* tubes, the preserved burrows of ancient worms.

Three quarters of a mile east of the south end of the bridge over the James River, the route crosses a fault that shoved Pedlar granulite gneiss onto Chilhowee sandstones. Near the interchange with the Blue Ridge Parkway, the route crosses the Rockfish Valley fault, which brings granulite gneiss of the Lovingston massif onto the rocks of the Pedlar massif.

View to the west of the James River gorge through the Blue Ridge

On the west side of Lynchburg, the route passes from granites of the Lovingston massif to the overlying Moneta gneiss. This thoroughly recrystallized gneiss may be sheared and recrystallized Grenville basement rock or it may be a sedimentary unit between basement and overlying Lynchburg group.

Blue Ridge Parkway James River — Chestnut Ridge

56 mi./90 km.

South of the James River, Parkway roadcuts reveal both layered granulite gneisses and massive, coarse-grained charnockitic granites. Some of the banding in the gneisses is not apparent on fresh dark green surfaces but stands out notably on the tan weathered surfaces. Some of the gray streaks are strung out quartz grains; others are dark crystals of mica.

The Peaks of Otter are held up by massive charnockite, a pyroxene granite, of the Peaks of Otter suite of rocks. These rocks are part of the Pedlar massif of the Blue Ridge geologic province. The gray-green charnockite is exposed in the road cut across from and a few dozen yards south of the lodge.

Southwest of Peaks of Otter, the Parkway swings northwest and

Sharp Top, one of the Peaks of Otter, is held up by massive charnockite.

Bedding planes of Chilhowee sandstone form a roadcut at milepost 97 on the Blue Ridge Parkway.

goes around the Glade Creek window through the Blue Ridge thrust sheet. The Blue Ridge thrust fault was arched here at the end of the Alleghanian orogeny, then eroded through. From Five Oaks Overlook, you can look down into the window and see a surface eroded onto rocks of the early Paleozoic carbonate bank. Hills in the thrust sheet across the valley and above the carbonates are much older Precambrian granites and gneisses similar to those between the overlook and the Peaks of Otter.

A roadcut across the Parkway from Five Oaks Overlook exposes dark gray-green granulite gneiss. Notice that the layering is nearly horizontal but that a dike cutting the rock is nearly vertical. This outcrop is near the Precambrian surface upon which the Chilhowee sandstones were deposited beginning in Eocambrian times. Less than a mile from this overlook the road climbs the stratigraphic section into the Chilhowee sandstones.

Between mileposts 91 and 101, several overlooks provide views of the James River valley. The near ridges with tan quarries cut in them are the Chilhowee sandstone foothills to the Blue Ridge. The nose off to the right across the river is Purgatory Mountain, an anticline of Tuscarora sandstone with barren outcrops on its flanks.

The Pulaski thrust fault wraps around the nose of Purgatory Mountain and stays on the northeast side of the James River for several miles. Movement on the Pulaski thrust fault placed rocks of

Massive beds of Chilhowee sandstone weather spheroidally at milepost 95 on the Blue Ridge Parkway.

the early Paleozoic carbonate bank on top of younger Silurian formations, including the Tuscarora sandstone. Further deformation then folded the fault. The Pulaski fault arched over the top of Purgatory Mountain before it was eroded away. All this folding and faulting are part of the Alleghanian deformation.

Look down into the quarry at the Quarry Overlook at milepost 101. Those trucks you may see on the far bench are Euclids with wheels 6 feet high. This quarry, originally opened to build the railroad, produces limestone and dolomite for construction aggregate. The rocks are part of the early Paleozoic carbonate bank caught in a fault slice below the Blue Ridge thrust fault, but above the Pulaski thrust fault.

South of the Quarry Overlook, the Blue Ridge Parkway descends into the valley of Goose Creek. At its intersection with US 460 the Parkway is on Cambrian rocks of the carbonate bank. Here erosion cut through the gently dipping Blue Ridge thrust sheet to open the Glade Creek window into the younger rocks beneath. The thrust fault carried billion-year-old granites and gneisses northwest over 500-million-year-old limestones and dolomites of the early Paleozoic carbonate bank. The nearly horizontal fault plane itself was later folded. Erosion then cut away the high parts of the thrust sheet to expose the younger carbonate rock beneath. The mountains you see to north, east, and south are twice as old as the rock underfoot.

The mountain you see from the Parkway south of the interchange with US 460 is Read Mountain, an anticline with Silurian Keefer sandstone in its core. The Pulaski fault runs near its base, but is folded over it and has eroded away above the mountain. Read Mountain is, thus, a window in the Pulaski thrust plate.

Near milepost 108, the Parkway approaches the Blue Ridge fault and then runs parallel to it for another mile before crossing onto the thrust sheet. The overthrust rock is a layered granulite gneiss, some much resembling granite. Several outcrops of it are in roadcuts between mileposts 100 and 111.

The road crosses the Rockfish Valley fault near milepost 112. Although the rocks on both sides of the fault are 1000-million-year-old Grenville granitic gneisses, this fault separates the two major divisions of the Blue Ridge province. It begins in northern Virginia in the core of the Blue Ridge anticlinorium and extends southwestward. Roadcuts near the Roanoke River expose spectacular fist-sized crystals of feldspar in granite.

Chilhowee sandstones make blocky roadcuts atop the northern Blue Ridge.

Quarries in the Arvonia formation have produced slate since Colonial time. James River in the distance
Photo courtesy of Arvonia-Buckingham Slate Company

US 15
James River — North Carolina state line
102 mi./164 km.

The highway crosses the James River at Bremo, where resistant sandstones in the Arvonia synclinorium tower above town. Bluffs on either side of the river are carved from both limbs of a small syncline within the synclinorium. Rare fossils indicate that this sand shoal was laid down in late Ordovician to early Silurian time, about 435 million years ago.

Arvonia is on a narrow anticline of Chopawamsic formation. Two modern quarries work the slate west of the highway; many others were worked in the past. Watch for the slate roofs and signs on buildings throughout the area. This rock has been in production since Colonial times and is currently shipped throughout the region for roofs, signs, wall facing, and patio flags. This slate is considerably harder than typical slate and, therefore, is not extensively carved.

Between Arvonia and Farmville, the highway is on high ground held up by resistant rocks of the Arvonia synclinorium. Views of the rolling Piedmont on either side are plentiful. At US 60, the Blue Ridge forms the westward skyline some 50 miles away.

Along US 15, from US 60 to US 460, are glimpses through the trees of the rocky crags of Willis Mountain, a ridge of quartzite in the Arvonia formation that contains abundant crystals of blue kyanite. Some zones are nearly pure kyanite in crystals somewhat larger than a thumb. Although most of the kyanite is processed for furnace brick, its most familiar use is in the ceramic of automobile spark plugs. Willis Mountain consists of two ridges that meet in an angle open to the northeast. Both ridges are mined. The highway crosses the structure at the junction of the ridges, so you cannot see the open part.

South of Willis Mountain is the Farmville basin. The Triassic and Jurassic sediments that fill this basin include small coal measures, which were worked in the last century. South of Farmville the route crosses two small basins which may have connected to the Farmville basin before erosion.

US 29
Amherst — Danville
80 mi./129 km.

The highway crosses onto rocks of the Lynchburg group midway between Amherst and Lynchburg. Good exposures exist on either side of US 29 business at the north end of the bridge over the James River. Metamorphosed sedimentary and igneous rock formations are exposed there along River Road.

The Lynchburg formations were deposited on the North American continental slope of the Iapetus, or ancestral Atlantic, Ocean. Water circulation must have been restricted in that ancient ocean basin because these formations typically contain graphite and pyrite. Both minerals indicate a lack of oxygen on the sea floor. Many of the Lynchburg sedimentary units are interbedded with dark volcanic rocks and intruded by masses of dark igneous rocks. Lenses of sandstone and marble appear toward the top of the group.

South of Lynchburg the highway crosses the southwestern nose of Candler Mountain. Bedrock there is a slice of the Candler formation, phyllite and schist interbedded with sandstone and some marble. A thin fault slice of Lynchburg group lies east of Candler Mountain.

The mineral turquoise has been valued for its color from time immemorial. Almost never, however, does it grow crystal faces. In the now abandoned Bishop Mine west of Lynch Station, however, pinhead-sized blue to greenish crystals line fractures in the gneiss.

The route angles across the Smith River allochthon between the Bowens Creek and Chatham faults. The Smith River allochthon is a slice of Inner Piedmont rock that was shoved over the Sauratown Mountains during the Taconic orogeny in late Ordovician time. The Leatherwood granite, which intrudes the allochthon, crystallized during late Ordovician time.

Crushed and deformed gneisses of the Fork Mountain formation in the Chatham fault zone east of Chatham contain the largest uranium-ore body yet discovered in eastern North America. Pitchblende in this deposit yields more than four pounds of uranium oxide for each ton of ore and reserves are estimated at 30 million pounds of the oxide.

The Danville basin lies southeast of the Chatham fault for more than 100 miles in Virginia and North Carolina. Swamps in this basin laid down deposits of peat during Triassic time, while the Atlantic Ocean was beginning to open. Then the peat was covered by later sediments and turned into coal which has been mined for local use. Like many of the Triassic basins, faulting was greater on the northwest side than on the southeast.

South of White Oak Mountain, which is carved from a sandstone unit in the basin, US 29 crosses the Vandola fault onto the Shelton granite gneiss. It forms the riffles in the Dan River where the highway crosses it in Danville.

The Jackson River cuts across the strike of the ridges between Covington and Clifton Forge.

US 220
Clifton Forge — Roanoke
43 mi./69 km.

Between Covington and Clifton Forge, US 220 follows the Jackson River Valley through Devonian shales between the Warm Springs anticline to the northwest and the Rich Patch anticline to the south. Lower Devonian Helderberg limestone forms the cliff at the east end of the bridge over the Jackson River. Just east of the bridge is a small cave on the north side of the river in the roadcut for old US 60.

Abandoned iron mines at the town of Low Moore were active from the middle of the nineteenth century to the first quarter of the twentieth. This valley was then the state's leading producer of iron ore. The ores were limonite that weathered out of the limestones. The deposits were too small to compete with the Minnesota ores after the first world war.

Much of the shale exposed in roadcuts along Interstate 64 is greatly contorted and fractured. This intense deformation may mean that a fault lies nearly flat below the surfaces. Shale taken from the highway cuts was used as fill between the roadcuts. Ground water then oxidized the pyrite in the shale to produce sulfuric acid, which promptly dissolved the calcite. That causes the shale fill to compact and buckle the highway, which requires repeated maintenance.

Near Clifton Forge a large water gap, Rainbow Gorge, comes into view to the south. The Jackson River cut through the Rich Patch anticline to give a spectacular view of the structure. Resistant sandstone ledges outline the structure, a fold overturned to the northwest with gently dipping layers on its southeast limb. The upper arch is the Eagle Rock sandstone and the inner arch is the Tuscarora sandstone.

Between Rainbow Gorge and Eagle Rock Gorge, the route passes along a shale valley cut into gently folded Devonian strata. Just north of Eagle Rock, the Pulaski fault rises to the surface, but its exact location in the shales is unknown. The James River cut its spectacular gorge through two blocks of middle Paleozoic rock separated by a fault. First the eastern block was shoved up the Pulaski thrust fault. Then the western block was thrust over the eastern block. Finally, both blocks were overturned to the northwest. Now they are separated from the Devonian shales by one branch of the Pulaski fault and from the Cambrian limestones to the south by another branch.

South of Eagle Rock, the route climbs out of the James River valley onto rocks of Ordovician age. At Fincastle, a spectacular conglomerate appears in roadcuts just north of the town. South from Fincastle is Tinker Mountain, the north end of a syncline with Mississippian Price sandstone in its trough. The route crosses a broad, flat synclinal downfold in rocks of the early Paleozoic carbonate bank between Fincastle and Cloverdale.

Eagle Rock sandstone forms the roadcut along US 220 at Eagle Rock on the James River. Photo by Thomas M. Gathright, II. Courtesy of the Virginia Division of Mineral Resources

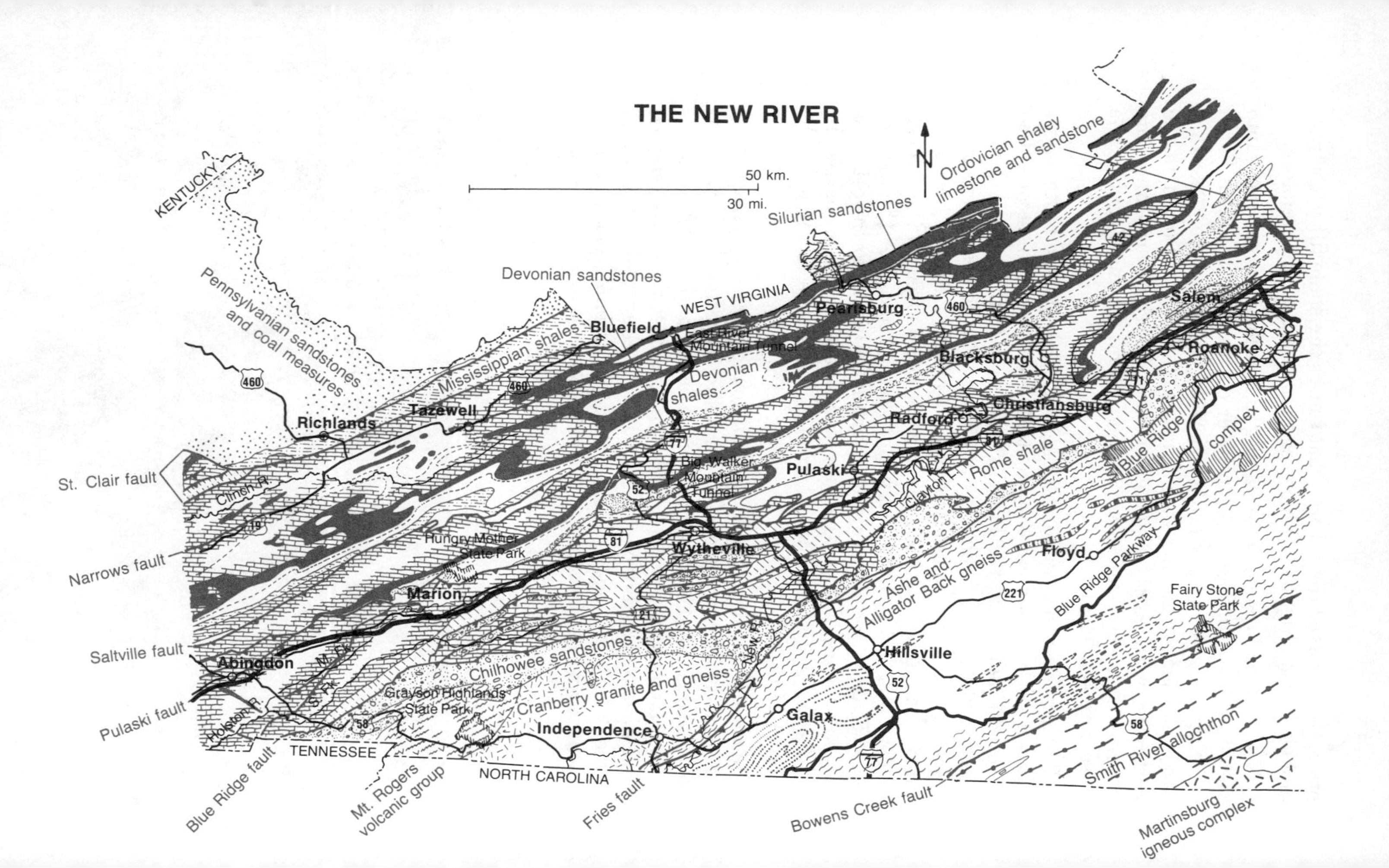
THE NEW RIVER
50 km.
30 mi.
KENTUCKY
WEST VIRGINIA
TENNESSEE
NORTH CAROLINA
Ordovician shaley limestone and sandstone
Silurian sandstones
Devonian sandstones
Devonian shales
Mississippian shales
Pennsylvanian sandstones and coal measures
Rome shale
Blue Ridge complex
Ashe and Alligator Back gneiss
Chilhowee sandstones
Cranberry granite and gneiss
Smith River allochthon
Martinsburg igneous complex
St. Clair fault
Narrows fault
Saltville fault
Pulaski fault
Blue Ridge fault
Mt. Rogers volcanic group
Fries fault
Bowens Creek fault
Richlands
Tazewell
Bluefield
East River Mountain Tunnel
Big Walker Mountain Tunnel
Pearisburg
Blacksburg
Christiansburg
Radford
Pulaski
Salem
Roanoke
Wytheville
Marion
Abingdon
Hungry Mother State Park
Grayson Highlands State Park
Fairy Stone State Park
Independence
Galax
Hillsville
Floyd
Blue Ridge Parkway
Clinch R.
Holston R.
New R.
Claytor L.
460
19
77
52
81
21
58
221
42
11

VII
THE NEW RIVER
ACROSS AN ANCIENT LANDSCAPE

WHAT'S NEW ABOUT THE NEW RIVER?

Headwaters of this most unusual river are in northwestern North Carolina near the Tennessee state line. From there, the North and South forks flow generally eastward then northward to their confluence three miles south of Mouth of Wilson. The New River follows a generally easterly course that takes it twice back into North Carolina before it meanders northeasterly across the Blue Ridge province to its gorge cut in Chilhowee sandstones between Cold Ridge and Poplar Camp Mountain. Virtually the entire Blue Ridge province from Mount Rogers northeastward to the escarpment leading to the Roanoke basin drains into the New River or its tributaries.

The New River continues its meandering course diagonally across the great carbonate valley of the Valley and Ridge province to a gorge in the Clinch sandstone between Cloyds and Gap mountains. In the valley, its tributaries extend from Wytheville to Christiansburg.

North of its gorge through the sandstone, the New River takes a northwesterly course across the western Valley and Ridge, cutting spectacular cliffs through each resistant bed that it encounters. Its course displays a cavalier disregard for rock erodability as the channel cuts through hard and soft rock alike. For a short distance north of the town of Narrows, however, its channel is cut into the easily erodible breccias in the Narrows fault zone.

The New River continues its meandering course across the flat-lying formations of the Appalachian Plateaus province to its confluence with the Gauley River south of Charlestown, West Virginia, where it becomes the Kanawha River, a tributary to the Ohio. The New is the southernmost river on the North American continent that flows from south to north and the only river to cut its course through the entire Appalachian Mountains from southeast to northwest.

Topography of the New River drainage basin in the Blue Ridge province is a gently rolling upland bounded by steep scarps of the topographic Blue Ridge to the southeast and the Roanoke basin to the northeast. Quartz boulders in terraces high above the present channels near the Blue Ridge escarpment indicate the former existence of an extensive drainage area to the southeast, now drained by streams that flow into the Atlantic Ocean. Wind gaps in ridges on these uplands, permit geologists to estimate that the New River has eroded a minimum of 4000 feet of rock.

Cliffs of Knox dolomite (early Paleozoic carbonate bank) rise above the New River in near Eggleston. Photo by Thomas M. Gathright, II; Courtesy of the Virginia Division of Mineral Resources.

The meandering course of the New River across the grain of the Valley and Ridge province is one line of evidence that suggests the New River course was established in Paleozoic times. Clearly, its size and erosive power were sufficient to permit it to maintain its channel even as the Appalachian chain was being folded during the Alleghanian orogeny.

But if an orogeny did not divert the New River, the Pleistocene ice ages did. The Ohio River downstream from the Kanawha is a new river. As glacial ice blocked northward drainage and choked old valleys with till, rain and melt water flowed along the front of the glacier and carved their channels where they could. Thus, water from the New River was diverted down the new Ohio River to the Mississippi and eventually to the Gulf of Mexico.

Where did the New River empty before the ice ages? Analysis of the shapes of the bluffs above the Ohio River between the Kanawha and Pittsburgh reveal that the valley was originally cut by a stream flowing to the northwest. From Pittsburgh, the ancestral New River possibly flowed northward into the St. Lawrence River. Before the opening of the Atlantic Ocean, it may have dumped its load into Cretaceous seas that invaded the North American continent.

A quick look at a globe will reveal that diversion from an ancient northern course to the Ohio and Mississippi system greatly shortened its distance to sea level. Shortened, the gradient steepened permitting it to deepen its meandering channel as can be seen along much of its course.

What's new about the New River? The present New River is but a fragment of a once mighty system. And it is not new at all, but older than the mountains it traverses—perhaps, the oldest river on the North American continent.

WHY THE SOUTHERN BLUE RIDGE?

In sharp contrast to the Blue Ridge north of Roanoke, that south of there to the North Carolina line is an erosional escarpment. East of the crest of the topographic Blue Ridge, drainage is into the Atlantic Ocean by way of the Roanoke River to the north and the Peedee River near the North Carolina state line. West of the crest, water collects to the New River and flows into the Gulf of Mexico by way of the Ohio and Mississippi rivers.

Since the waters here have a shorter distance to sea level in the Atlantic than in the Gulf of Mexico, the stream gradients are much steeper to the east than to the west. Steeper gradients mean that the streams have a higher erosive power to the east. So the divide between Atlantic and Gulf drainage slowly shifts to the west with every storm that breaks on the Blue Ridge.

Could we travel backwards in time, we would find longer tributaries to the New River and shorter tributaries to the Roanoke and Peedee rivers. Indeed, could we travel back 200 million years to a time before the opening of the Atlantic Ocean, we might find the entire area to the east draining northwestward into a mighty ancestor of the New River. More recently, we would find the New River shrinking in size as the Atlantic drainage nibbles away at the headwaters of the New River by a process called stream piracy.

Although the topographic Blue Ridge is a line making a drainage divide, the geologic province of the same name is a broad upland extending northwest to the Blue Ridge fault which marks the boundary with the Valley and Ridge province. In the Piedmont southeast of this divide the rocks are no different from those immediately to the northwest. Erosion carved the Piedmont surface after the Atlantic Ocean opened, beginning in Triassic time. Much of the upland surface, by contrast, may be little changed since Pangaea assembled in the late Paleozoic.

To the southwest, the schists, gneisses, and greenstones are named for their exposures in North Carolina, the Ashe formation below and the Alligator Back above. To the northeast is the similar Lynchburg group which geologists think correlates with the Ashe. Detailed mapping in the central part that could establish this correlation remains to be done. But if correlations established in central Virginia can correctly be extended to the southwest, the contact between the Ashe and the Alligator Back may be a yet-unreported thrust fault between the two formations.

Mount Rogers from Elk Garden.

MOUNT ROGERS NATIONAL RECREATION AREA

Not until detailed mapping of the area in the nineteenth century, did anyone realize that Mount Rogers is the highest point in Virginia. Other mountains, such as the Peaks of Otter near Roanoke, stand out in sharper relief from their surrounding lowlands. Mount Rogers, however, is perched on already high ground. Climbing Mount Rogers is no more difficult than hiking the Appalachian Trail.

White Top Mountain, adjacent to Mount Rogers and the second highest peak in the state, has a public road going nearly to its summit. The summit itself is not open to the public because of a microwave relay station there. From the public parking loop on clear days you can look out across the Iron Mountains to Walker and Clinch mountains in the distant Valley and Ridge province.

Bedrock holding up Mount Rogers is the Mount Rogers group; they are distinctly different from just about any other rocks in the state, but vaguely similar to formations cropping out in the Grandfather Mountain window in North Carolina.

The Mount Rogers formation is divided into three member units. The oldest consists mostly of basalt lava flows with subordinate amounts of rhyolite volcanic ash and sandstone. The middle member, making up about half the total thickness, is predominantly rhyolite with small amounts of basalt and sediments. The youngest member is predominantly sediments with small amounts of basalt and rhyolite. Pebbly to bouldery siltstones and a variety of conglomerates crop out in the very youngest horizons. Some of the siltstones contain isolated pebbles and boulders.

Sediments in the Mount Rogers formation are generally gray to green in the earliest part of the sequence and maroon in the latest part. Rocks of the entire formation have been mildly metamorphosed.

Several hypotheses have been advanced to account for these enigmatic rocks. Rafting by floating ice could explain how boulders get into the muds that became siltstones. Glaciers could account for the piles of rock rubble that became conglomerates. But the late Precambrian equator was too close for glaciers to form.

Mysterious boulders are observed today on the silty sediments of the playas of desert mountain basins of the American West and many such basins are ringed with storm-generated debris flows. The rocks of Mount Rogers may have formed in

This conglomerate of the Mount Rogers formation has been interpreted as a Precambrian glacial till or debris flow.

This laminated siltstone with graded bedding has been interpreted as glacial varves or turbidity current deposition.

such a semi-arid basin ringed by hills of granite and some volcanoes. Quiet lakes might have been periodically racked by violent storms that generated debris flows and surges of sediment.

Another controversy about Mount Rogers is its place in the geologic column compared with other formations in Virginia. Some geologists think the base of the formation was deposited on the underlying Grenville Cranberry gneiss and that Chilhowee sandstone was deposited on the Mount Rogers group. Since its rocks are so unlike others of similar age, some geologists argue that it is allochthonous, that it was transported along faults from somewhere else. This, of course, merely moves the problem elsewhere. Until further study of the area resolves the question of the origin of the Mount Rogers group of formations, it must remain a suspect terrane.

HOW LONG WILL THAT LAKE LAST?

Lakes are ephemeral features of the Earth's crust, slowly, but surely, destroying themselves from the day they form. Some self-destruct faster than others. Man-made lakes are of the faster variety.

Of all the lakes that dot the Virginia landscape, only two are natural—Lake Drummond in Chesapeake and Mountain Lake in Giles County. And of these two today, Lake Drummond is more artificial than natural, being maintained by the United States Army Corps of Engineers by a series of locks.

Usually a little exploration and common sense can reveal how any particular lake came into being. Examples include Crater Lake in the collapsed crater of a dormant volcano in Oregon, glacial lakes in the Midwest, oxbow lakes in cut-off meanders along the Mississippi River, and sinkhole lakes in collapsed caverns in Florida.

The origins of Virginia's two natural lakes, however, have sparked controversy among experts for years. Lake Drummond is on a terrace of unconsolidated sediments and by all logic and reasoning should have eroded itself a drainage years ago. Some of the ideas put forth for its origin range from meteorite impact to swamp fire, but its origin remains an enigma.

Mountain Lake, as the name implies, is on top of a mountain, a mountain that is certainly not a volcano. Here again, ideas for its origin run from meteorite impact to limestone cavern collapse to cows stomping around a salt lick on top of a mountain.

Recent geologic investigations reveal that the bottom contours of Mountain Lake are much like those of an artificially dammed river, but the dam in this case is a mass of blocks of Clinch sandstone. A perfectly ordinary mountain stream was eroding its way headward through a ridge held up, like many in the Valley and Ridge province, by the Clinch sandstone. Undercutting less resistant formations underneath caused blocks of this sandstone to break off and slide downslope to dam Mountain Lake.

Other lakes in Virginia are man-made or misnamed bodies of water. But natural or man-made, lakes eventually fill up with sediment or erode through their spillway. Some natural lakes have been around for tens of thousands of years; some artificial lakes have filled completely while others have burst their dams.

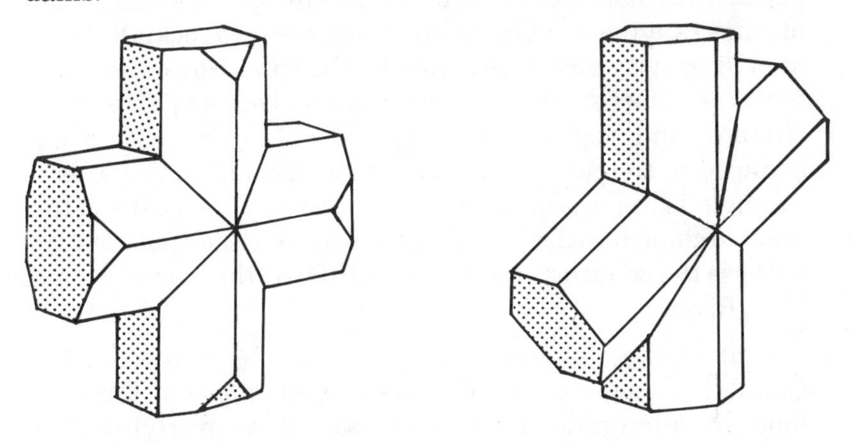

Staurolite crystals commonly form twins, crossing at 90 degrees as on the left or at 60 degrees as on the right.

FAIRY CROSSES

Staurolite is a perfectly ordinary mineral found in some metamorphic schists and gneisses. It is an iron aluminum hydroxyl silicate and commonly grows as crystals in rocks of appropriate composition and grade of metamorphism. Typically these crystals are in the form of elongate prisms, but more often than not they grow as twins.

Crystal twins are sort of like Siamese twins; certain portions are shared, a desirable trait in crystal twins. Twinning is actually very common in the mineral kingdom, but staurolite twins take the form of crosses and weather out of the rock. When the mineral was formally named in 1792, the mineralogist took the Greek words *stauros*, cross, and *lithos*, stone, and created the name staurolite. Some twins cross at right angles, others at 60 degrees.

Early Virginia settlers found these twin crystals and called them fairy stones. Fairy Stone State Park, near Martinsville, is just one locality where you can find staurolite twins.

SLOPES STABLE AND UNSTABLE

The only stable slope is a nearly horizontal slope. Material on an inclined slope—hillside or mountainside—moves downhill at rates ranging from inches-per-year creep to rockfall accelerated by gravity and slowed only by the wind. Slope instability, however, depends on such variables as bedrock, slope angle, climate, and vegetation. Some slopes appear not to have changed in the memory of the oldest resident in the area; an example is the profile of Stony Man on Skyline Drive. Others seem to change with almost every casual passage past the area; witness the countless repairs of US 460 at the Narrows on the New River.

Well off the highway and south of the New River in Giles County, geologists recently discovered monster landslides more than ten miles long. There, sandstone overlying shales dip gently toward a valley. The river cut down through the sandstone, thus removing a major prop keeping the sandstone

Unstable slopes fail in a variety of ways. During slump, one or more slices slides down and out from a slope along a curved fracture with a backward sense of rotation so that many trees or posts are tilted uphill.

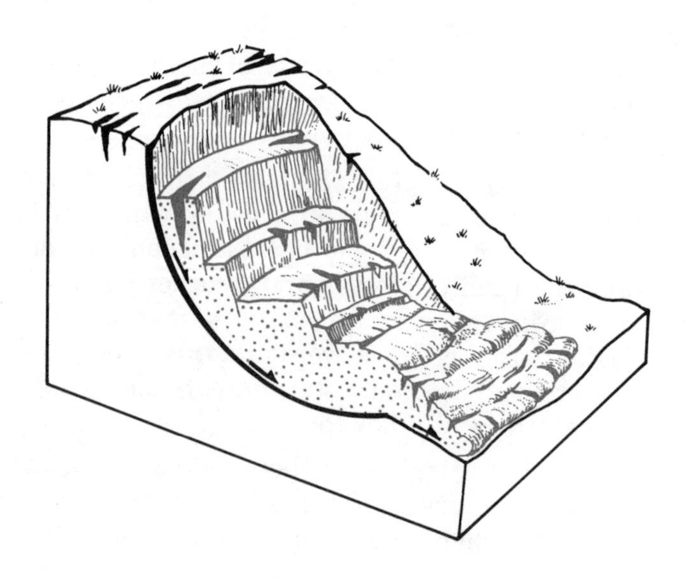

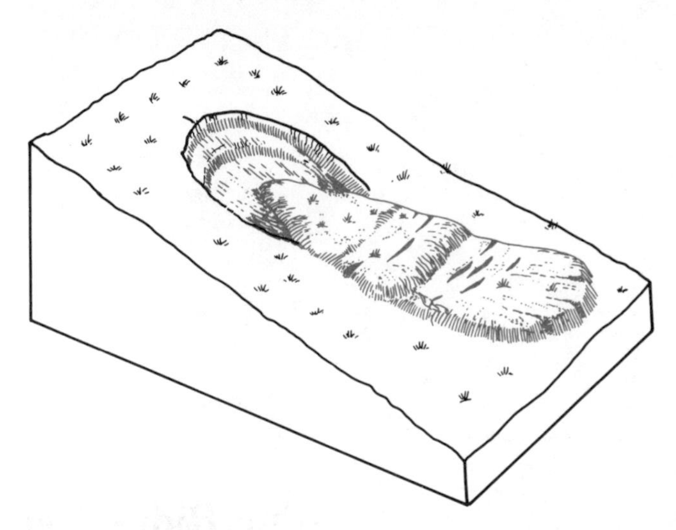

During a debris slide, rock and soil flow out and down the slope. Many slumps develop debris slides at their toes. Failed slopes are shown in color. You can recognize an old failed slope by its chaotically tipped trees and shrubs. Since once-failed slopes can be reactivated, they are not good places to build roads or houses.

in place on the hillside. In some unknown, prehistoric time these sandstones slid toward the valley creating thousands of acres of chaotic terrain. No one in their right geologic mind would develop a mountain resort on this shifting land.

At Narrows, fault-crushed rock was undercut by the New River eroding its channel and banks. Then came the railroad along the valley needing space for its tracks and cutting deeper into an already unstable slope. Highway engineers further undercut the slope to put in the roadbed for US 460. You can see the resulting bare slide scarps and feel the ever heaving highway as you drive by there. And you can see the buckled steel retainer below the highway if you stop and look down.

Blue Ridge Parkway
Chestnut Ridge — North Carolina state line
96 mi./154 km.

Between mileposts 120 and 130 there are many good overviews of the Roanoke valley. The most spectacular is near milepost 130. Both the Blue Ridge to the north and the ridges of the Valley and Ridge province stand out in stark relief. On exceptionally clear days you can see the Rich Patch Mountains some 50 miles to the north.

From the Metz Run Overlook near milepost 128 to milepost 131 there are several roadcuts in an extremely coarse grained granite gneiss. Fist-sized feldspar crystals are set in a matrix of dark mica, feldspar, and quartz. This rock is quite distinct from the banded gneiss that crops out between mileposts 136 and 138.

The climb from the Roanoke Valley is from the erosional surface of the Piedmont, which here extends through the Blue Ridge province into the Valley and Ridge province, to the erosional surface of the New River valley and its tributaries. The New River is, without a doubt, the most misnamed river in the state of Virginia. Geologic evidence indicates that it had established its course by Mississippian times, 350 million years ago. By contrast, no Atlantic drainage could have existed before the beginning of its rifting in Triassic times, 200 million years ago.

The Roanoke River has captured large portions of the New River drainage. As evidence, the South Fork of the Roanoke River flows northward for ten miles before joining the North Fork flowing in an easterly direction toward Roanoke. The North Fork flows southwest for ten miles before turning eastward. These initial directions of the two forks of the Roanoke River are more appropriate to tributaries of the New River, which flows in a northerly direction 15 miles to the west and some 400 feet higher in elevation.

Beginning about milepost 133, the character of the Blue Ridge becomes distinctly different from that north of Roanoke. To the north it is a single ridge flanked on the west by ridges of Chilhowee sandstone and on the east by deeply dissected granitic rock terrain. Both sides drain into the streams which ultimately discharge into the Atlantic Ocean. From here south it is an erosional escarpment.

The Parkway is near the divide between the lower Piedmont drainage, which flows eastward into the Atlantic Ocean, and the higher New River drainage, which flows westward into the Ohio and Mississippi rivers, ultimately into the Gulf of Mexico. Because the New River system is several hundred feet higher than the Piedmont, erosion by streams flowing east from the divide is the more active. Tributaries of the New River are periodically captured by stream piracy as the divide slowly shifts westward.

Views to the southeast and to the north contrast sharply. To the southeast you see a broad panorama of the rolling Piedmont terrain below, broken by isolated erosional remnants such as Cahas Mountain and Bull Mountain. To the northwest, although the terrain is gently rolling, vistas are not so broad because the Parkway is but little higher than these tributary valleys to the New River.

Sometime in the next few million years, the Roanoke River will

Some phases of the charnockite consist of fist-sized feldspar crystals set in a matrix of feldspar, quartz, and pyroxene. These roadcuts are near the gorge of the Roanoke River through the Blue Ridge.

As the mica schist deformed, quartz layers necked down to form a string of sausage shapes (boudinage) near milepost 162 on the Parkway.

behead the New River, perhaps in the vicinity of Radford, leaving the broad Blue Ridge uplands subject to erosional attack from both sides as is the northern Blue Ridge. Old uplands will be weathered and eroded away and old rivers like the New will suffer successive decapitation of their headwaters.

Near Devils Backbone, the Parkway passes from rocks of the Blue Ridge basement complex onto the metamorphosed oceanic sediments and volcanic deposits of the Lynchburg group. A good exposure of these rocks is across from Rakes Millpond at milepost 162. Here, deformation of the thinly layered gneiss to black schist stretched veins of quartz into a series of boudin, sausage-shaped segments. Under pressure the host rock could flow but the rigid quartz veins broke apart.

Although the Parkway continues on rocks of the Lynchburg group, deep weathering of the schists and gneisses prevents their exposure in many roadcuts. Notable exceptions are at Rocky Knob, milepost 172-176, Mabry Mill, Groundhog Mountain, Fancy Gap, and milepost 216 on Virginia 89. North of Meadows of Dan, the Parkway passes onto the Alligator Back formation.

To the south the Blue Ridge divides Atlantic Ocean drainage to the southeast from Gulf of Mexico drainage to the northwest. The province is a broad, undulating surface extending from the topographic Blue Ridge on the southeast to the Blue Ridge thrust fault on the northwest.

The horizontal distance to sea level in the Gulf is more than five times that to the Atlantic. Therefore, river systems that drain to the Atlantic are steeper and more erosive than those that drain to the Gulf. South of milepost 130, the Parkway follows the divide between Atlantic and Gulf drainage.

Streams draining northwest from the Parkway are in many cases in large valleys indicating original erosion by a stream much larger than is there today, even after periods of heavy rain. These valleys were cut when the Blue Ridge front was southeast of its present location giving them bigger drainage catchment basins and larger stream flow.

These now underfit streams had their headwaters decapitated by more vigorous streams, a process called stream piracy. Examples of recent stream piracy where Atlantic streams captured former New River drainage are at Rocky Knob Park and Meadows of Dan. Abundant exposures of rocks of the Lynchburg group at Rocky Knob result from rapid erosion at the headwaters of Rock Castle Creek, a tributary of the Smith River. At Meadows of Dan, the Dan River beheaded Reed Island Creek, a tributary of the New River, so recently that erosion by the Dan River has not had time to establish the expected steep eastward course.

Mabry Mill near the crest of the Blue Ridge near Meadows of Dan.

The Presbyterian Church at Willis is made of vein quartz collected from fields of the area.

US 221
Roanoke — Hillsville
71 mi./114 km.

The route begins in Roanoke in the Roanoke River valley on the Cambrian Rome formation in the lower part of the early Paleozoic carbonate bank. South of Cave Spring, it crosses the Blue Ridge thrust fault to the Grenville granites and granite gneisses of the Pedlar massif. Meanwhile, the road climbs out of the Roanoke Valley and onto New River drainage. Between Airport and Adney Gap, the drainage was pirated by headward eroding tributaries of the Roanoke River system.

Between Copper Hill and Check, US 221 crosses from Grenville granite gneiss to the overlying late Precambrian Lynchburg group of metamorphosed sedimentary rocks. Here the topography becomes distinctly more angular because the layers of rock differ in their resistance to erosion. In almost every field along the road, you can see large chunks of white vein quartz.

North of Floyd, watch for outcrops in the fields with layers that dip to the southeast at 40 to 50 degrees. The trend of the layers is roughly parallel to the road. Geologists name this rock the Little River gneiss, but are uncertain whether it formed through metamorphism of igneous or sedimentary rocks. The gneiss contains distinctive augen of pale feldspar about the size of your thumb and the quartz grains are blue.

Some downtown buildings in Floyd are made of soapstone bricks that were locally quarried. At Willis, the Presbyterian church is made of vein quartz float that weathered out of the bedrock. You can see that local farmers have devoted considerable time in removing this hazard from their fields, piling it along the edges or centers of their fields.

Close up of vein quartz in church wall.

US 11, Interstate 81
Roanoke — Bristol
148 mi./238 km.

South of Roanoke, US 11 climbs out of the Roanoke valley and crosses an asymmetric divide to the watershed of the New River valley east of Christiansburg. The divide is asymmetric because the slopes toward the Roanoke drainage were much steeper than those to the New River drainage. Many spectacular roadcuts were blasted out of early Paleozoic bank limestones between Salem and Christiansburg on the north side of this divide.

Because the streams are so much steeper on the Roanoke side than on the New River side, erosion is faster to the north than to the south. So the divide shifts south as the more vigorous Roanoke streams capture parts of former New River streams. Thus, the very old New River drainage to the west shrinks as the relatively young Atlantic drainage to the east encroaches upon it.

The New River is entrenched here as indicated by its meandering narrow floodplain. Geologists are not sure of the cause for the geologically recent downcutting of the channel, but some speculate that Pleistocene glaciation diverted it from a Canadian discharge to its present discharge in the Ohio River. Such a diversion would shorten its distance to the sea, thereby increasing its erosive power by steepening its gradient.

At Christiansburg both routes cross the Christiansburg window through the Pulaski thrust sheet, exposing limestones of the early Paleozoic bank underneath the older Cambrian clastics. At the New River, Interstate 81 crosses the Claytor thrust fault into Cambrian

Devonian shales in roadcuts on the west side of Draper Mountain south of Pulaski.

limestones and then the Max Meadows thrust fault into older Cambrian Rome siltstones east of Draper Mountain.

Between Christiansburg and Radford, US 11 crosses several splays of the Salem fault. One slice brings Devonian shales to the surface. Just west of the river, the highway crosses the Salem fault onto early Paleozoic bank limestones and follows them to Dublin. Between Dublin and Pulaski it crosses into Cambrian Elbrook formation. Movement along the Pulaski fault fractured the lower rock in the Pulaski thrust sheet. Watch for broken rock where Peppers Ferry Road runs into US 11.

South of Pulaski, US 11 crosses the Pulaski thrust sheet onto Mississippian and Devonian shales, which crop out abundantly on the north side of Draper Mountain. The structure is an anticline. The spine of the mountain is the Tuscarora sandstone. The view is spectacular from the crest across the New River valley to Macks and Poplar Camp mountains underlain by Chilhowee sandstones.

The route runs on early Cambrian formations most of the way between Draper Mountain and Wytheville. Hills to the south, Lick Mountain, are made of slivers and slices of Chilhowee sandstone. To the north is Cove Mountain, eroded in Tuscarora sandstone. Wytheville is on early Paleozoic bank limestones at the east end of an syncline that has Ordovician sedimentary rocks in its core.

Between Wytheville and Marion, the routes run along the Max Meadows thrust fault. South of the highways you can see Lick Mountain south of Wytheville, then Glade Mountain, and Pond Mountain

View across New River valley from Draper Mountain on US 11.

south of Marion. Each is a series of half a dozen or so fault slices of Chilhowee sandstones reared up toward the sky. Between Lick and Glade mountains you can see through to the Iron Mountains beyond. The Iron Mountains are carved from a single slice of Chilhowee sandstone bounded on the north and south by thrust faults.

Between Marion and Bristol, the routes continue more or less along the trend of the rocks of the early Paleozoic carbonate bank on the Pulaski thrust sheet. To the northeast, Walker Mountain in the Saltville thrust sheet is held up by Cambrian Copper Ridge dolomite. To the southeast, the Iron Mountains are held up by Chilhowee sandstones.

Just north of Bristol, the interstate swings around to the west and crosses onto the thrust sheet above the Bristol fault. This thrust sheet appears to have been folded and then slid into place from the southwest during Alleghanian deformation. The interstate parallels this fault less than a quarter of a mile northwest from the Interstate 381 interchange to the Tennessee state line.

US 460
Christiansburg — Claypool Hill
107 mi./172 km.

The interchange between Interstate 81 and US 11/460 is in the Christiansburg window eroded through the Pulaski thrust plate into the older rock beneath. The window reveals an anticline in underlying Cambrian strata of the early Paleozoic bank. The route crosses out of this window onto the Pulaski thrust sheet and then onto the Price Mountain window south of Blacksburg. There, a bulge in the thrust sheet permitted erosion to expose younger Mississippian rocks beneath the fault. Price Mountain near the center of the window is Mississippian rocks poking through much older Cambrian rocks.

The Mississippian sedimentary rocks exposed in a borrow pit behind the Blacksburg animal hospital are the youngest in Virginia outside the Coastal Plain. They are relatively undeformed, whereas the Cambrian limestones that were shoved over them are contorted almost beyond recognition.

Between Blacksburg and Brush Mountain, the road crosses the Mississippian Price formation, which contains the oldest mineable coal seams in the state. According to local legend, the largest coal seam was named the Merrimac because it supplied coal to the ironclad ship of that name during the Civil War.

The valley between Brush and Sinking Creek mountains is etched into easily erodible Devonian shales, mudstones, and sandstones. Sinking Creek Mountain is held up by Silurian formations, chiefly the Clinch sandstone. The line between Giles and Montgomery coun-

Silurian Rose Hill sandstone reveals mudcracks outlined with hematite at the crest of Gap Mountain along US 460.

ties is close to the boundary between Ordovician Martinsburg formation to the west and Silurian sandstones to the east. Mud cracks outlined in red hematite here show that the deposition site was periodically above water. The Appalachian Trail follows this sandstone ridge.

Sandstone layers stretched into sausage shapes in the Martinsburg formation.

East of Pembroke, the road follows a limestone valley along the flank of Spruce Run Mountain, a sliver of Clinch sandstone caught up as a slice between branches of the Saltville fault. Sinking Creek valley and Virginia 42 follow one branch of the Saltville fault. This fault has little displacement here, but farther southwest it becomes a major structure in the Valley and Ridge province.

Tomahawk Hill in Pembroke is the result of the New River cutting off a meander before the channel reached its present elevation. Stream gravels in terraces a couple of hundred feet above the present channel contain vein quartz pebbles which could have come from no nearer than the Blue Ridge, some 50 miles to the southeast.

Pembroke is near the north end of the Bane dome. In the center of the Bane dome, Ordovician fossils were extracted from rocks that were underneath older rocks of Cambrian age. These relationships indicate that the Narrows-St. Clair thrust sheet was domed, and has been eroded through to younger rocks below.

The US 460 bypass around Pearisburg crosses the trough of a syncline near the west end of the bridge over the New River. Two large limestone quarries appear north of the bridge. The nearer one was abandoned after a quarry wall weakened by solution of the limestone dropped on the equipment. More than 40 species of Pleistocene and recent mammal, reptile, and snail fossils have been recovered from the caves and sinks in this quarry.

From Pearisburg to Narrows, the route crosses and recrosses Devonian shales and Cambrian limestones in a series of fault slices. The trace of the Narrows fault crosses the road near the west side of the Celanese plant in the town of Narrows. Rocks in the Narrows fault are extremely broken and hence, easily eroded. The New River follows the fault for a short distance. On the north side, the river undercut the zone of breakage causing slumping and sliding.

Fault breccia descending upon US 460 in slumps and slides near Narrows. Note tilt of the power pole. White material visible through the trees is freshly disturbed ground.

The full extent of the Narrows landslide can be seen from fields across the New River.

Cuts for the railroad and highway caused more sliding. Watch for the freshly bared ground above the highway near the eastern entrance to the gorge. The first recorded movement of this slide was in 1916, when it ruined a section of the railroad. Major and continuing failure of this slope began in 1940, following modernization of the highway. When the highway was widened to four lanes in 1970, slippage increased causing two of the lanes to be closed temporarily. In spite of the money spent for retaining walls and the removal of 250,000 cubic yards of rubble, the tipped poles, fresh excavation, and bent steel in the retaining wall on the south side of the highway, all indicate recent movement of this unstable slope.

The New River gorge between Narrows and Rich Creek passes through Peters (East River) Mountain which, like so many ridges in the Valley and Ridge province, is capped by the Clinch sandstone. Layers of sandstone dip steeply in the gorge and cause the rapids in the river.

The gorge of the New River through the Narrows. The railroad bed is visible, but the highway is hidden in the trees.

The axis of the Hurricane Ridge syncline is the boundary between the Valley and Ridge and the Appalachian Plateaus provinces. To the southeast of the hammer, the beds become vertical. To the northwest the beds become horizontal.

The Saint Clair fault is exposed in the west-bound lane through the Narrows of the New River. Movement along the fault shoved Cambrian limestone over younger Devonian shale. This is one of the few places in southwestern Virginia where you can place your hand on a fault plane. Be careful though, because traffic is heavy and the berm extremely narrow. When the leaves are down, you can see this same fault in the railroad cut across the New River, where it includes a block that was detached at depth and trapped between two branches of the fault. Although the Saint Clair fault can be traced for hundreds of miles to the southwest, it dies out in an anticline just north of the Narrows.

Between Pembroke and the West Virginia state line, the highway follows the valley of the New River, the only one that cuts through the Valley and Ridge province from east to west. A few hundred yards north of the highway along highway 648 near the state line is a classic geologic locality, where the axis of the Hurricane Ridge syncline is exposed in the late Mississippian Hinton sandy siltstone. The axis of the fold is where the bedding planes turn from horizontal to upright and overturned within a distance of less than a foot. This exposure is the structural front of the Valley and Ridge province. The highway crosses the axis of the Hurricane Ridge syncline just west of the junction with US 219. Watch for it when the leaves are down from the east end of the bridge over the New River.

Bedding planes on the northwest side of US 460 just across the West Virginia state line are overturned to the northwest. What you see is a worm's-eye view of ripple marks and other sediment structures.

For about one mile southwest of the West Virginia state line, the roadcut on the northwest side of the highway exposes overturned bedding planes of the Mississippian Bluefield formation. The rock varies from limestone to limey shale. Because the bedding is overturned to the northwest, what you see is a worm's eye view of fossils and sedimentary structures. Analysis of sedimentary features, such as ripple marks in the rocks, indicates that they contain a sequence of cycles in which sea water became progressively shallower, then suddenly deeper.

Some geologists think these cycles may be related to cycles of glacial advance and melting in the southern hemisphere. Others suggest an internal cause on a slowly, but steadily subsiding continental shelf. They reason that growth and the production of limestone is much faster than subsidence, quickly filling the area to sea level and choking off further growth. According to this interpretation, growth does not then resume until the water regains a depth necessary to permit circulation to start anew. Then the area quickly fills and the cycle repeats.

Between the New River and Bluefield, the highway crosses spectacularly eroded, flat-lying strata of the Appalachian Plateaus. Ancient sand-filled channels exposed in some of the roadcuts mark where streams eroded their channels only to have them filled with later deposits. East of Bluefield, the highway recrosses the Saint

This death bed of quickly killed ancient animals may tell of excess saltiness or some other Paleozoic pollution.

Clair fault back onto the Valley and Ridge province.

Between Bluefield and Tazewell, East River Mountain stands prominently southeast of the highway. Northeast of Bluefield it marks the West Virginia state line and becomes the structural front of the Valley and Ridge province. Midway between Tazewell and Bluefield limestone outcrops in the roadcuts have bedding planes that dip toward the synclinal mountain. These limestones are part of the early Paleozoic carbonate bank; the road follows Bluestone River roughly along their trend.

The Narrows fault disappears into a syncline of Ordovician rock near Tazewell and displacement along it increases to the northeast. This displacement doubles the Silurian section with the result that both East River Mountain on its northwest side and Buckhorn Mountain on its southeast side are Clinch sandstone. The Narrows fault, unlike other thrust faults in the Valley and Ridge, dies out to the southwest, probably because displacement was taken up by the Saint Clair fault, the next thrust fault to the northwest.

On the north side of the west-bound lane on the Tazewell bypass bedding planes in the limestone dip toward the road. Watch for ripple marks and mud cracks on some of these surfaces. These indicate that the formations were deposited near sea level and exposed to the

atmosphere during low tides. These rocks are used as retaining walls in the town of Tazewell.

Limestone outcrops in the roadcuts between the Tazewell turnoff and Claypool Hill show the effects of solution by ground water. Freshly blasted rock is gray but the weathered solution channels are tan. Many of these solution surfaces are studded with knobs of chert as big as grapes.

East River Mountain to the left is the West Virginia state line from Bluefield almost to the New River Narrows.

Spectacular cliffs in Silurian sandstones were created for Interstate 77 through Wolf Creek Mountain.

US 52, Interstate 77 Bluefield — North Carolina state line

69 mi./111 km.

This route enters the state inside East Mountain tunnel under North Gap east of Bluefield, West Virginia. This ridge, like so many others in the Valley and Ridge province, is held up by the resistant Clinch sandstone. Here the structure is a narrow syncline. The Narrows fault brings Cambrian carbonates onto younger Silurian formations on the south side of the fold.

The forks of Laurel Creek run approximately along the axis of this syncline before joining at North Gap to flow southward through the gap. Outcrops here expose Devonian formations. Between North Gap and Rocky Gap the route follows the narrow gorge of Laurel Creek through Buckhorn Mountain with excellent exposures of Silurian sandstones on both sides. At Rocky Gap, the Narrows fault brings Cambrian limestone against younger Silurian sandstones.

Wolf Creek Mountain is a ridge of Clinch sandstone which is well exposed in the roadcuts. South of this ridge the dips of the Devonian strata become much gentler than farther north as the route crosses a

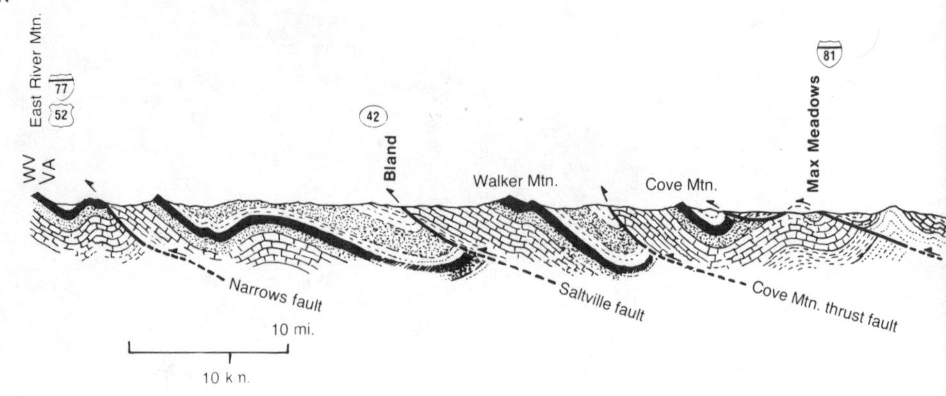

Cross section of Valley and Ridge, Blue Ridge, and Piedmont geologic provinces along Interstate 77 from West Virginia to North Carolina. Valley

broad anticline cored by shaley formations. Round Mountain west of Bastian is an anticline folded in Clinch sandstone. Farther southwest erosion has cut into the fold to open a circular valley almost entirely surrounded by a ridge of the sandstone, Burkes Garden. Northeast on this fold, erosion has cut through the Narrows thrust fault into older Ordovician rocks in the sheet below.

The broad ridge between Bastian and Bland is Brushy Mountain, capped by the Mississippian Price sandstone, another major ridge former in the Valley and Ridge. A small thrust fault is exposed in Devonian shale in the truck run-out ramp for the northbound lane of the interstate highway. The big red sandstone cuts on either side of the route are in the Price formation.

A backthrust fault dipping to the north crops out through the truck run-out for Interstate 77 north on Brushy Mountain.

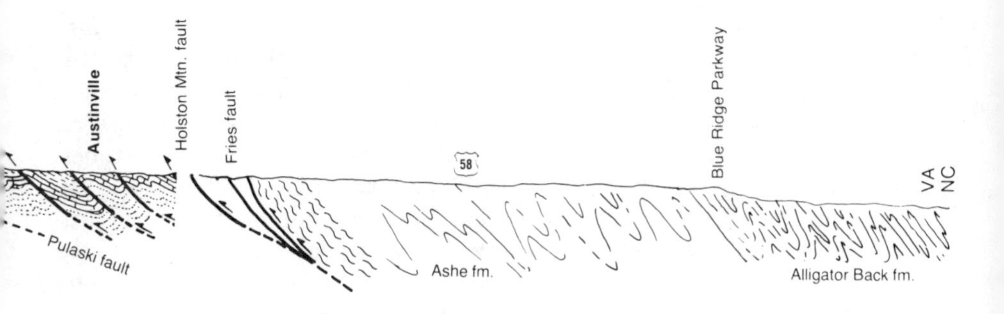

and Ridge extends from Bluefield to the Holston Mountain fault. Blue Ridge province extends from that fault to the Blue Ridge Parkway. Southeast of the Blue Ridge escarpment lies the Piedmont province.

At the town of Bland, the Saltville fault brings Cambrian Honaker limestone over Mississippian Price sandstone. Big Walker Mountain, immediately south of Bland, is another ridge held up by the Clinch sandstone. US 52 south of Bland winds over the top of the ridge some 500 feet above Walker Creek. Beyond that is the north wall of Burkes Garden. You can climb the tower or ride the lift for even broader views on a clear day.

Immediately south of Big Walker is Little Walker Mountain, another ridge of Price sandstone. On its south side, the Cove Mountain thrust fault brings up another slice of Clinch sandstone. Just south of Cove Mountain, the Pulaski thrust fault brings Cambrian Elbrook limestone to the surface. US 52 passes west of where the Pulaski fault has overridden the Cove Mountain fault.

Wytheville is in a small syncline in Cambrian and Ordovician limestone of the lower Paleozoic carbonate bank. Lick Mountain to the south is underlain by half a dozen or so fault slices of Chilhowee sandstone.

The broadly anticlinal structure of the Cambrian limestone south of Fort Chiswell is complicated by minor thrust faulting. These limestones crop out in some of the fields as the route approaches the New River. From the New River to Hillsville, the routes cut directly across a former mining district, the Gossan Lead ("leed"). Lead ("led") was extracted from sulfide ores in Colonial times around Austinville and Ivanhoe. Rifle balls for the Patriot troops were cast from this lead at Shot Tower on the bluff above the New River. Lead and zinc mining continued until the 1970s. Other mineral products from the Gossan Lead included copper, pyrite for iron and sulfuric acid, manganese, titanium, barite, kyanite, and soapstone. Exploratory drilling for gold in the 1970s failed to reveal commercial concentrations.

Poplar Camp Mountain is a slice of Chilhowee sandstones caught between faults. Near the town of Poplar Camp, the road crosses the Holston River fault. On the south side of the mountain, movement along the Stone Mountain thrust fault shoved gneisses of the Precambrian Ashe formation onto the younger Chilhowee sandstones. This thrust fault marks the boundary between the Valley and Ridge and the Blue Ridge provinces. Bedrock north of here is unmetamorphosed sedimentary rock, mostly deposited in ancient continental seas. From here south, the bedrock is moderately to strongly metamorphosed sedimentary and igneous rocks.

Less than a mile south of Poplar Camp Mountain, the Fries thrust fault brings in a sliver of Grenville Cranberry granite gneiss. This is succeeded immediately by gneisses of the late Precambrian Ashe formation, the bedrock all the way to Hillsville. The Ashe formation includes schists, gneisses, sandstones, and conglomerates. These rocks have been eroded into rounded hills almost without bedrock outcrops. This rolling topography extends almost without interuption from the Poplar Camp Mountains to the Blue Ridge front at Fancy Gap.

About a mile north of the Blue Ridge Parkway, gneisses of the Ashe formation are overlain to the south by gneisses of the Alligator Back formation. Geologists think these formations correlate with the Lynchburg and Candler groups, respectively, in central Virginia.

The Blue Ridge south of Roanoke to the North Carolina state line is an Atlantic-facing erosional escarpment. The arrow shows where the Blue Ridge escarpment was a few million years ago. Or you can say that the arrow points to the present scarp and the black lines indicate slope retreat expected in the next few million years.

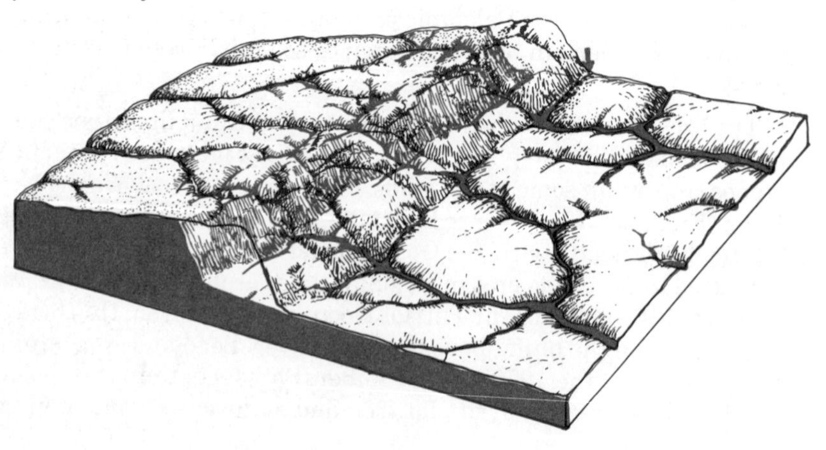

At Fancy Gap, the routes cross the Blue Ridge Parkway. From the vantage here, the panorama of the Piedmont stretches south and east to the Blue Ridge front. This front is an erosional escarpment where Piedmont streams cut away at the divide that separates their headwaters from those of the tributaries to the New River. No major change in rock type marks this boundary between the Blue Ridge and Piedmont geologic provinces.

US 21
Wytheville — North Carolina state line
36 mi./58 km.

Wytheville is at the east end of a syncline in rocks of the early Paleozoic carbonate bank. Just south of the city, the route crosses the Max Meadows thrust fault onto the lower Cambrian Rome formation. It continues through progressively older rocks toward Lick Mountain.

Lick Mountain contains a half dozen or so thrust slices, each repeating a sequence of Chilhowee sandstones. Valleys open into the mountain where dolomite caught between slices of sandstone has allowed faster erosion. Glade Mountain, visible to the west, has a similar structure. A small cavern in that roadcut on the south side of Speedwell is in the same dolomite, the lowest member of the early Paleozoic carbonate bank.

US 21 runs past a small, undeveloped cavern on the south side of Speedwell. Note the six-inch stalactites.

A freshly cleaved surface from a block of Striped Rock granite in an abandoned quarry just north of Independence reveals clots of black mica and hornblende in feldspar and quartz.

About a mile south of Speedwell, at the foot of the Iron Mountains, the Holston Mountain fault shoved Chilhowee sandstones onto the younger formations of the carbonate bank. Although most roadcuts are covered with greenery, massive, red, vertically dipping beds of Chilhowee sandstone are visible just south of the Mount Rogers National Recreational Area on the south side of the mountain. A pebble conglomerate bed in the Chilhowee group is exposed about 100 yards down the hill from the massive sandstone.

South of the park boundary near the junction with highway 805, the Stone Mountain fault thrusts Grenville Cranberry granite gneiss onto the Chilhowee sandstones. Look in the roadcut in the ridge south of Elk Creek, about 7½ miles north of Independence, for Cranberry granite and granitic gneiss. These rocks are cut by late gabbro dikes which enclose fragments ripped out of the Cranberry granite.

Fields and hillsides along the valley north of Independence are littered with boulders. About two miles north of Independence is the contact between the Cranberry granite gneiss and the late Precambrian Striped Rock granite. This rock is named for the broad stripes of black stain you see on the bare exposures west of the highway just north of Independence.

Views north from Independence are dominated by exposures of Striped Rock granite. To the south is the deeply etched terrain of the New River valley. The southern contact between the intrusive Striped Rock granite and the host Cranberry gneiss is about a block south of the intersection with US 58 and US 21.

About 3 miles south of US 58 the route crosses the Fries thrust fault, a major feature dividing the Blue Ridge province. Here the fault appears to thrust gneisses of the Ashe formation onto the Cranberry granite gneiss. Schists of the Ashe formation are in the roadcut near the North Carolina state line on the south side of the New River.

View near the North Carolina state line downstream to the east of the New River which eventually runs to the northwest into West Virginia.

Three-inch staurolite crystals in this roadcut in Galax are a bit surprising. They are prisms, not the cruciform twins found at other localities.

US 58
Meadows of Dan — Abingdon
115 mi./185 km.

Meadows of Dan is at the top of the Blue Ridge escarpment, on the boundary between the Blue Ridge and Piedmont geologic provinces. Bedrock here is mica schists and gneisses called the Alligator Back formation. West of the Blue Ridge Parkway, the road passes onto mica gneiss of the underlying Ashe formation. Geologists do not yet agree whether the contact between Ashe and Alligator Back formations is depositional or a thrust fault bringing Alligator Back rocks in from the southeast.

An unusually coarse phase of the Ashe formation is exposed in the roadcut on the north side of US 58 and 221 at the traffic light at bypass Virginia 89 in Galax. Here you may find thumb-sized crystals of staurolite which, unlike the Virginia fairystone crosses of this mineral, are untwinned. You may also find smaller crystals of garnet, apatite, and tourmaline, as well as large crystals of mica.

Between Galax and Independence, toward the west end of the

stretch of road parallel to the New River, Precambrian gneisses of the Ashe formation lap onto older Cranberry gneiss. West of the river stretch, the road crosses a series of slivers of the Fries fault, a major thrust fault that splits the Blue Ridge geologic province.

At Independence the route crosses a corner of the outcrop area of the Striped Rock granite. You can see bald knobs of this granite with its wide stripes of black on light gray on the mountainsides north of town. For about a half mile west of Independence, US 58 runs close to the contact between Striped Rock granite on the north and Cranberry gneiss, a metamorphosed granite. Round boulders of Cranberry granite gneiss litter the fields between Independence and Mouth of Wilson.

About one mile north of Volney on Virginia 16, the Cranberry granite gneiss is exposed in large pavements east of the road. A few hundred feet north of a small store there, a rhyolite dike intruded the granite gneiss. The mineral grains are large enough to be visible in the center of the dike, but are too fine to see near the contact with the granite gneiss, where the rhyolite cooled very quickly. You can get another good look at the Cranberry granite gneiss in the massive roadcut at Rugby, where it weathers to a pale tan color and shows little sign of metamorphic foliation layering.

Two miles west of the turn-off toward Rugby the route crosses from Cranberry gneiss to the Mount Rogers volcanic rocks. Roadcuts between Cabin and Laurel creeks expose pale gray volcanic rock—rhyolite—and conglomerate. Some of the rhyolites contain a few pebbles of granite about the size of walnuts that were ripped off the volcanic pipe during its eruption. Mount Rogers is held up by volcanic rocks, chiefly rhyolite.

This granite boulder was dropped onto a silty bottom during deposition of the Mount Rogers formation.

Waterfall over boulders of conglomerate in Mount Rogers National Recreation Area.

The highway department has excavated an interesting small quarry in the metamorphosed mudstones on the east side of highway 755 and Laurel Creek just north of US 58. Search the steeply inclined bedding planes of these mudstones for isolated granite boulders dropped there when the mud was still fresh. These dropstones have puzzled geologists for decades. Proposed origins for them have ranged from chunks of granite blasted out by a volcanic eruption to rafting by ice which subsequently melted and let them sink to the muddy bottom of an ancient lake or sea. They remain an enigma.

Between Laurel Creek and Straight Branch, the road crosses the Stone Mountain thrust fault onto Chilhowee sandstones. Whether or not the fault marks the actual boundary between Mount Rogers formations and the Chilhowee formations is a question as yet unresolved among geologists.

At Damascus, the road crosses the Iron Mountain thrust fault, which moved sandstones of the Chilhowee group over formations of the lower Paleozoic carbonate bank. Rocks in the roadcuts east of Damascus were probably deposited scores of miles to the southeast before they were shoved into their present locations.

Folds and small faults in these shallow marine carbonates control the topography between Damascus and Interstate 81. The Middle and South forks of the Holston River here follow the weaker strata of infolds of middle and upper Ordovician rocks; the carbonates form the ridges and many of the roadcuts.

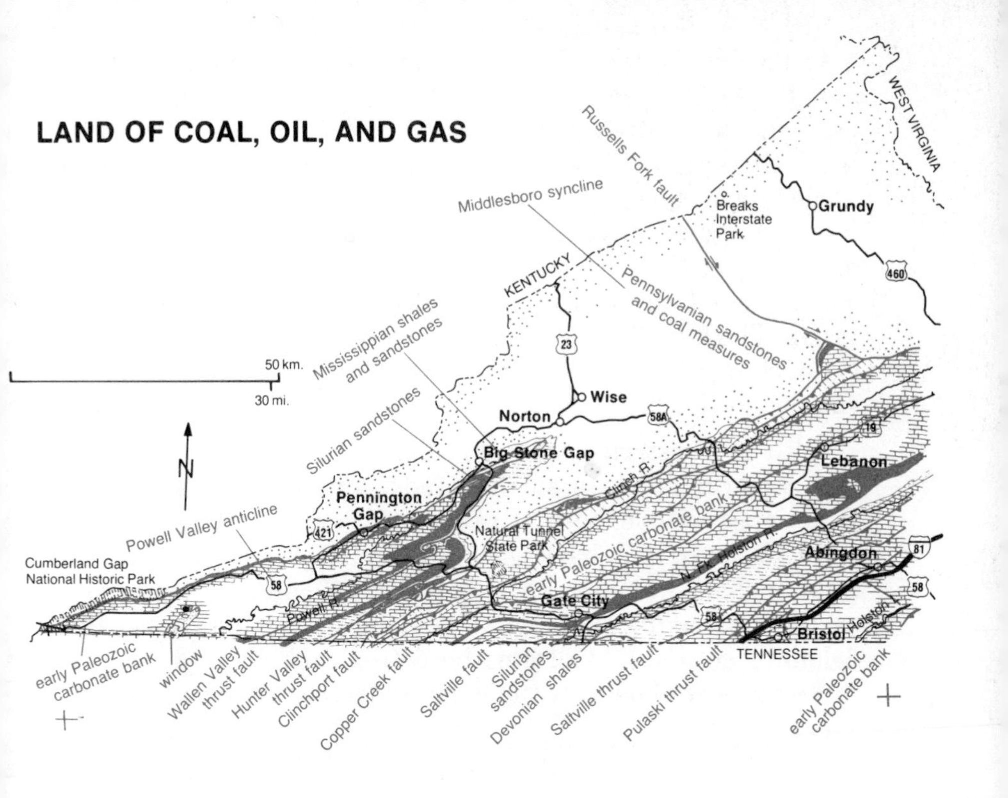

VIII
LAND OF COAL, OIL, AND GAS

THE VALLEY AND RIDGE AND THE APPALACHIAN PLATEAUS

Wells drilled in the Appalachian Plateaus province show that it has the same rocks in the subsurface as appear at the surface in the Valley and Ridge. The overwhelming difference is that rocks of the Valley and Ridge were folded and faulted by the Alleghanian orogeny at the end of the Paleozoic Era whereas those of the plateaus were not. Until that happened, their geologic histories were largely the same.

In late Precambrian time, the area that is now the Valley and Ridge and the Appalachian Plateaus provinces was an eroding surface of older igneous and metamorphic rocks. Rifting to the east marking the opening of the Iapetus Ocean left an ancient shoreline somewhere near the eastern edge of the Valley and Ridge province. Beaches and off-shore sand shoals accumulated there as sediments washed off the continent to the west. These sediments became the sandstones of the Chilhowee group. The later Chilhowee sandstones contain fossils of early Cambrian age and the entire group of formations may be early Cambrian.

With the western source of sediments eroded away, a shallow, tropical sea covered the sands; limestone and dolomite were deposited over the entire area. Although there were local sand beaches and bars, mostly they were much like the Bahama bank of today. For more than 100 million years, from early Cambrian to middle Ordovician time, limestone and dolomite accumulated from these tropical seas and became those rocks we now call the early Paleozoic carbonate bank.

One of the early formations of the carbonate bank, the early to middle Cambrian Rome formation, contains enough silt and clay to make parts of it very shaley. These shaley layers deform easily, so that much fault displacement occurred by slippage within the Rome formation when the Valley and Ridge province was folded and faulted at the end of Paleozoic time.

The massive sandstone block that dropped off the roadcut from above the coal seam is about six feet high.

Carbonate deposition ceased for several million years during middle Ordovician time. Sea level dropped. Carbonates deposited in early Ordovician time exposed to rain water and solution produced a ragged karst topography on the low plain.

Then came the Taconic orogeny in late Ordovician time, around 450 million years ago. Rocks formed in the Iapetus Ocean were shoved onto the North American continent above the carbonate bank to become the Taconic Ranges of eastern New York state. The Inner Piedmont belt of the Carolinas appears to have docked on the North American continent at this time. With it, the Smith River allochthon was shoved over the Sauratown Mountains and onto the eastern part of the Blue Ridge province. This doubling of crustal thickness would produce the highlands that shed erosional debris onto the suddenly deepened carbonate bank.

In the area to become the Valley and Ridge province the effect was to deepen the seas to depths beyond where sunlight could penetrate. Sands and muds poured into this basin from the east where the Taconic orogeny had heaved up mountains near the continental margin. The deepest part of this basin received sediments only periodically in the form of great surges of turbid, muddy water, possibly churned up by monster storms. These middle Paleozoic shales and sandstones became a second sequence of weak rocks that would slide easily during later deformation.

West and southwest of Virginia, however, the water remained shallow and clear enough for carbonate deposition to continue.

Gradually the basin filled with sediments and was topped off with sands mixed with the iron oxide mineral, hematite. These became the Juniata redbeds, some of them with mudcracks that show deposition within the tidal range and periodic drying by the atmosphere. On top of these red sandstones came a great shoal of pure quartz sand which became the Clinch sandstone, known as the Tuscarora sandstone to the northeast, as the Massanutten sandstone in the Shenandoah Valley. Many of the prominent ridges of the Valley and Ridge province are held up by these sandstones.

Then warm, shallow seas returned, permitting more carbonates to accumulate. This middle Paleozoic carbonate bank

In open folds in this Devonian shale, the southeast flank of the syncline and the northwest flank of the anticline are steeper than the common central flank. This is typical of folds in the Valley and Ridge geologic province.

lasted but 30 million years from late Silurian to early Devonian time. Then came the Acadian orogeny and a second deepening of the continental sea. Another chain of volcanic islands or microcontinents docked on North America, perhaps the extension of the Piedmont province now hidden beneath the sediments of the Coastal Plain.

Into the trough between the continental interior and the newly raised highlands to the east came the sands and muds that became the late Devonian shales of many of the valleys in the Valley and Ridge province. As the basin again filled by early Mississippian times, muds gave way to another great sand shoal. These, now the Price and Pocono sandstones, form prominent cliffs and ridges on the western margin of the Valley and Ridge and over much of the Appalachian Plateaus. The Price formation also contains the oldest mineable coal seams in Virginia.

Shallow, warm seas returned to these provinces and, by middle Mississippian times, carbonates were again accumulating. Vast swamps growing on that coastal plain laid down deep deposits of mucky black peat. Periodic rise in sea level flooded these swamps and buried the peat. It later became the coal seams now mined in the Plateaus province in the southwestern corner of the state.

Coal-producing cycles ended with the Alleghanian orogeny,

final closure of the Iapetus Ocean, and the assembly of Pangaea. Rock formations of the Valley and Ridge province were shoved from the southeast toward the northwest. They responded by folding into great anticlines. Some of the anticlines broke on their northwest sides to become thrust faults. Later, even the nearly horizontal thrust faults were caught up in the folding and were themselves overturned to the northwest.

The southern Appalachians, south of the Roanoke bend in the trend of the ridges, differ from the central Appalachians in important ways. The number of thrust faults is much larger and increases to the southwest. Many faults in the southern Appalachians die out in anticlines in the general zone of the Roanoke bend. The amount of transport across each fault or fault system increases to the southwest. Synclines lie between faults rather than between anticlines, as is the case north of the Roanoke bend.

Beginning in Triassic time, the modern Atlantic Ocean and Gulf of Mexico began to open as the New and Old Worlds drifted apart. With a new ocean to the east and south, the new continental margin thinned as the continent split. Erosion carried away the sediments and carved much of the present topography. Rock formations of the Valley and Ridge are contorted as a result of the Alleghanian orogeny whereas the formations of the Allegheny Plateau still lie almost as flat as they were laid down.

Interbedded limestone (white) and shale (gray) were crumpled during the Alleghanian orogeny in this roadcut on old US 460 southwest of Tazewell.
Photo by Photo by Thomas M. Gathright, II; Courtesy of the Virginia Division of Mineral Resources.

WHY THE VALLEYS AND RIDGES?

Mountains and valleys characterize the western third of Virginia. What is a mountain? Topographically, a mountain stands well above the surrounding land. Most, though not all, mountains are grouped into chains. The Appalachian Mountain chain extends from Alabama to the Maritime Provinces of Canada. Before opening of the Atlantic Ocean split the continents, the chain continued through the Caledonides of Scotland and Norway.

In the Appalachian chain, Precambrian and Paleozoic rocks have been folded and thrust, one stack upon another, from southeast to northwest. The most common Paleozoic rocks are sandstones, shales, and carbonates. Carbonates are dissolved by ground water, forming cave systems which eventually collapse, thus lowering the ground level. Shales, many of them cemented by calcite, erode easily and are washed away by streams and rivers. Sandstones, especially if cemented with quartz or hematite, neither dissolve nor easily erode.

In the Valley and Ridge province, the valleys are underlain

The Valley and Ridge geologic province contains folded sedimentary strata of the Paleozoic. Anticlines which close at the top alternate with synclines that close at the bottom. The ridges are held up by weather-resistant sandstone (solid black).

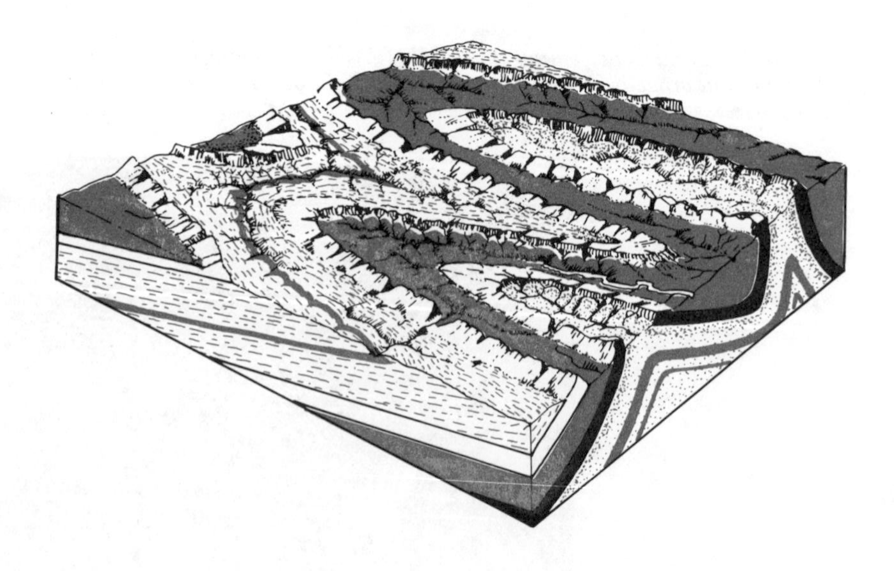

by nonresistant shale and carbonate rock whereas the ridges are supported mainly by resistant sandstone. Most are held up by one of two extremely resistant sandstone formations—the Silurian Clinch, Massanutten, and Tuscarora sandstones, the Pennsylvanian Price and Pocono sandstones.

THIN SKINNED TECTONICS

Two related structures dominate the geology of the far southwestern corner of Virginia—the Powell Valley anticline and the Middlesboro syncline. These two folds in the Pine Mountain overthrust sheet have suggested to some geologists for nearly a century that the folds in the Appalachian Mountain range do not involve the crystalline basement rocks, but are confined to the layered sedimentary rocks that cover the basement. This is the idea of thin-skinned tectonics. Drilling and seismic exploration during the last two decades have confirmed the concept. Indeed, where thrust sheets are stacked, a fold in the upper sheet does not extend into the one below.

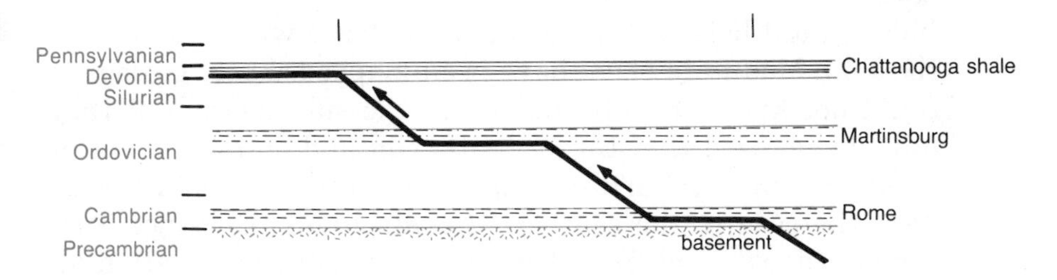

Thrust faults are composed of décollements *within weak beds and ramps through strong ones. The three main zones of decollement in the Appalachians are Cambrian Rome, Ordovician Martinsburg, and Devonian Chattanooga formations.*

Faulting begins in master décollement—literally an ungluing or unsticking—where strata above slide over strata below along a bedding-plane fault. The master décollement in the Appalachians is the shaley Rome formation of early to middle Cambrian age. Where deflected upward, the fault breaks through the rigid strata of the lower Paleozoic carbonate bank to the next higher weak formation, here the shaley beds of the

Martinsburg formation of late Ordovician age. The next ramp is through the rigid sandstone and carbonate formations of Silurian to Devonian age to the weak Devonian shales above.

As the moving upper plate goes from décollement to a ramp, it is flexed upward. When it goes from ramp to next higher décollement, it is flexed downward. Where the upper sheet drapes over the upper end of a ramp, it buckles into a rootless anticline that extends downward only to the fault.

The Powell Valley anticline, which extends for some 50 miles southwest from Norton into Tennessee, is just such a rootless anticline lifted when the Pine Mountain thrust fault ramped upward from one formation to the Martinsburg formation. Northwest of the Powell Valley anticline the Pine Mountain thrust fault ramps upward in the geologic section to the surface. The Middlesboro syncline lies between the anticlines.

Were this the whole story, the question of thin-skinned tectonics might not have surfaced until drilling and seismic exploration for oil and gas revealed it in the subsurface. But the structure of the Pine Mountain thrust sheet gives you windows through it to the rocks below. The Pine Mountain thrust warped upward under the Powell Valley anticline. That set the stage for the Powell River to cut through the fault plane to the hidden fault below. At present, half a dozen windows through the Pine Mountain sheet have been mapped. Early geologists could not know of faults hidden in the subsurface, but they quickly realized that thrust faults can and do flatten out at depth. Seismic and drilling exploration reveal that this is indeed the case in the southern Appalachians. Evidence from other windows and from drilling and seismic exploration all support the thin-skinned idea that thrust faults flatten out to the southeast and merge with a master décollement in the Rome formation. In the Powell valley, you can see the evidence.

Flat-lying late Paleozoic sandstones make dramatic roadcuts along US 58A.

REPEATING THE GEOLOGIC COLUMN

All roads cutting across the grain of the land south of the Roanoke bend in the Valley and Ridge geologic province cross a series of repetitions of Paleozoic strata. These repetitions are the work of thrust faults that shoved one slice of the stack of sedimentary strata up in a northwesterly direction upon the next stack.

Before the Alleghanian orogeny that ended deposition on the North American margin of the Iapetus Ocean, these strata rested in layer-cake fashion over a much wider area than they now occupy. How much wider? Some geologists estimate a shortening of as much as 50 percent in southwestern Virginia. By that reckoning, every 100 miles across the present deformed Valley and Ridge represents 200 miles of deposition on the continental shelf in the Paleozoic era.

At each of the major thrust faults, older Paleozoic rocks rest on younger Paleozoic rocks; in many cases Cambrian rocks rest on Mississippian rocks. Since these strata are tipped up at each thrust fault, erosion to the present level bevels the stack much like a curled-up telephone directory.

Lay your phone book face up on a table with its spine in a northeast to southwest direction. Fold the open edge up and you will have the names running from A to Z horizontally from southeast to northwest. Similarly, each fault slice has beds running from top to bottom as you travel from southeast to northwest across southwestern Virginia. Cross the next thrust fault to the next slice to the northwest and you begin anew much like having a series of telephone directories lying with their open edges overlapping the spine of the next one.

In this simplified model, each overlapping directory is a thrust slice. In southwestern Virginia each thrust slice is named for the thrust fault beneath it. Thus, the Pulaski sheet is that stack of strata lying above the Pulaski thrust fault and beneath the next major thrust fault to the southeast of it. In most cases, this is the Blue Ridge thrust fault.

Now, increase the amount of overlap to the southwest in your model to make it approximate more closely an early stage of the geologic structure of the southern Appalachian Valley and Ridge province where fault displacement generally increases to the southwest. Later in the Alleghanian orogeny, the thrust faults themselves were folded. Where subsequent erosion has carved through the arched part of the folded fault, you have a window through the upper slice to the lower one.

ROCKS FOR PROFIT

Small portions of the Earth's crust can yield large profits to people who own and work them. Metals can be extracted from metal-bearing minerals. These ores may be mined either from underground mines or from open pits. Other rock products may be nonmetallic, such as clay and limestone for Portland cement.

Metallic mineral deposits have been worked in Virginia since Colonial times. At various times Virginia has been a leading producer of iron, manganese, lead, zinc, gold, and titanium. No metals are now mined in the state, although the largest known deposit of uranium ore in eastern North America has been discovered northeast of Danville. As of this writing (1985) Virginia law prohibits mining uranium ore.

Nonmetallic mineral resources in Virginia fall into two categories—fossil fuels and minerals and rocks for construction and industry. Some 850 different mines produce in excess of 40 million tons of coal worth around $1.5 billion each year.

Bolts prevent the ceiling from falling at this coal mine. Photo by Thomas M. Gathright, II; Courtesy of the Virginia Division of Mineral Resources.

Most of the fossil fuel is in the far southwestern corner where coal, oil, and natural gas are produced. Gas is also produced in the western Valley and Ridge province. And the Triassic basins in the Piedmont province have produced minor amounts of coal.

In addition to the coal mines in the southwestern part of the state, at any given time some 250-300 mines, quarries, and borrow pits are in operation. Production is mainly from limestones of the Valley and Ridge, soapstone and granite from the Blue Ridge, granite and slate from the Piedmont, and sand and fill dirt from the Coastal Plain. The world's largest kyanite quarry is in Willis Mountain off US 15 between US 60 and US 460. Kyanite is an aluminum silicate mineral used in the manufacture of high-temperature brick for furnace liners and in the ceramic part of automobile spark plugs.

Another specialty mineral quarried in Virginia is vermiculite, a potash-deficient mica that expands 100-fold on heating. Expanded vermiculite is used for such diverse applications as light-weight concrete aggregate and as a major component of potting soils for gardeners.

Mineral extraction industries in Virginia produce only about one third the revenues produced by tourism, the state's leading industry. Only a handful of counties within the state, however, have no mineral extraction within their boundaries and in several it is a major source of employment and revenue.

Another way in which rock formations can be valuable is frequently overlooked by geologists and economic planners, alike. Some rock formations are eminently suited for disposal of waste products while others definitely are not. Limestones of the Valley and Ridge geologic province and unconsolidated sedimentary formations of the Coastal Plain geologic province are poor choices for water-soluble wastes which quickly wash through them from the disposal sites.

A court of law had to instruct the officials of one Virginia city not to park their urban trash on a parcel of real estate that was so riddled with caverns that nothing smaller than dogs would be filtered out. In subsequent drilling, 59 out of 60 test wells lost circulation before they reached their assigned depth; water is necessay for drilling and loss of circulation means you have drilled into a cavern. Tidewater's Mount Trashmores will be a blight on the Coastal Plain for centuries as leachate from the municipal refuse continues to flow into shallow aquifers.

Some shale formations of the Valley and Ridge and some crystalline rocks of the Blue Ridge and Piedmont provinces provide excellent containment for those things we want to get rid of and isolate from any future contact with living things. Saprolite may or may not provide containment depending upon the rock type it comes from. A good waste disposal site is a valuable piece of property—a reverse mine, if you will.

Red sandstone cliffs tower above the Levisa Fork of the Big Sandy River at Breaks Interstate Park at the West Virginia state line.
Photo by Thomas M. Gathright, II; Courtesy of the Virginia Division of Mineral Resources.

US 460
Claypool Hill — Kentucky state line
46 mi./74 km.

Between Claypool Hill and Richlands, the road crosses the Saint Clair fault and the boundary between the Valley and Ridge and the Appalachian Plateaus provinces. Here Cambrian rocks are thrust on top of Mississippian formations of the Appalachian Plateaus. Bedding planes of these rocks dip steeply. Very quickly you cross a section through the entire early Paleozoic carbonate bank sequence from late Ordovician rocks that underlie the Clinch valley to lower Cambrian strata that abut the fault.

North Gap in Sandy Ridge is the drainage divide between Clinch River drainage into the Tennessee River and Levisa Fork drainage into the Ohio River by way of the Big Sandy River. Steeper stream gradients on the northwest side of this divide indicate that streams are eroding faster there than on the southeast side. Erosion gradually shifts the position of the ridge to the southeast, increasing the Ohio drainage area at the expense of the Clinch.

Between the divide and the West Virginia state line, flat-lying red sandstones of Pennsylvanian age form spectacular cliffs on either side of the Levisa Fork valley crammed with river, railroad, highway, dwellings, stores, gas stations, and coal-mine heads. Virtually every bit of level ground has been utilized in this valley; some mobile homes seem to hang over cliffs at both ends. The economy is based on coal and you can see several seams exposed in the roadcuts. Coal in some of the seams is deformed and contorted whereas the overlying unyielding sandstones are undeformed.

US 19
Claypool Hill — Hansonville
31 mi./50 km.

For the most part, this connector road follows the grain of the landscape along carbonate valleys between sandstone ridges. Southwest of Claypool Hill, this limestone is prominent in quarries along the highway; in some cases these have been excavated down to a bedding plane that dips toward the highway. Both freshly blasted and old dissolved surfaces make up the roadcuts.

Near the county line, the route abruptly jumps across a syncline, passing from its northwest to its southeast side. Southwest of this jog, the road follows a narrow valley eroded along the northeast end of the Copper Creek thrust fault, another of those with displacement increasing to the southwest, like a giant scissors. Watch on the northwest side of the road for rock broken during the faulting. Such breakage is typically most conspicuous in the rocks above the fault. Because the broken rock is so prominent here, the road is probably above the fault.

Northwest of the highway, sandstone cliffs on Horse and Barn Mountain, held up by infolds of Silurian Clinch sandstone, reveal the resistence of that formation to erosion. South of Lebanon the ridge is Clinch Mountain, again Clinch sandstone.

On the south side of Lebanon, the road crosses the Copper Creek thrust fault, which here brings the Cambrian Honaker dolomite onto Ordovician limestones in the syncline. The road between Lebanon and Hansonville climbs into younger sedimentary rocks of the Ordovician Knox group.

From Pound Gap, red highwalls of countless strip mines rise above the sea of green of the Middlesboro syncline.

US 23
Kentucky state line — Tennessee state line
69 mi./111 km.

Pull off US 23 at Pound Gap and walk to the edge of the level cleared area. Try to count the number of reddish sandstone cliffs that rise above the forested hills in the land below. Each cliff is the highwall of a strip mine dug for coal, mainstay for the economy of the area.

Pound Gap in Pine Mountain is in the Pennsylvanian Lee sandstone. Just northwest of the state line the Pine Mountain fault thrusts these and underlying rocks over the younger Pennsylvanian formations. The field of view is in the flat-bottomed Middlesboro syncline, which extends to the town of Norton. Sandstone in the gap is tilted up to the west.

As it descends Pine Mountain, the road travels the geologic section through progressively younger rocks to the coal-rich Wise formation. Coal seams up to 2 feet thick crop out on the Pound bypass. Watch for the channel filled with sand in the shale exposed on the east side of the highway.

The landscape here is reminiscent of the mesa country of the American southwest. It is typical of the Appalachian Plateaus geologic province in Virginia. Massive sandstones cap flat-topped hills underlain by easily erodable shale and coal. This sandstone is used for retaining walls in the town of Wise and elsewhere.

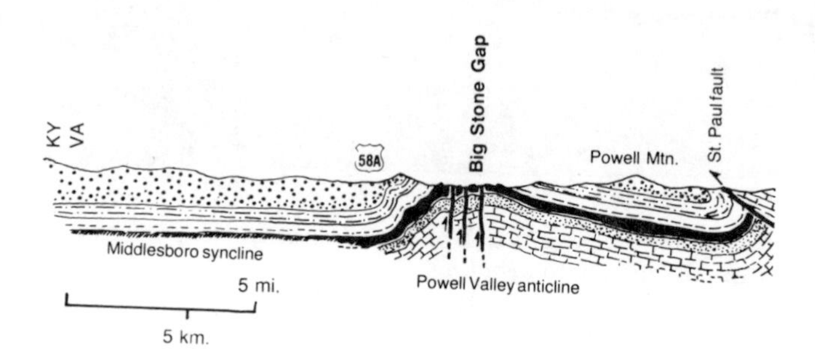

Cross section of Valley and Ridge geologic province

Appalachia is at the boundary between the Appalachian Plateaus and Valley and Ridge provinces. Here Stone Mountain is held up by many fault slices of Clinch sandstone. Bedding planes of the rock exposed in the roadcuts are steeply tilted.

At Jasper the road swings east across the fault, then across an anticline. The road travels through a gap in Powell Mountain in the

Mile-long Natural Tunnel off US 23 is large enough to admit a river and a standard gauge railroad.

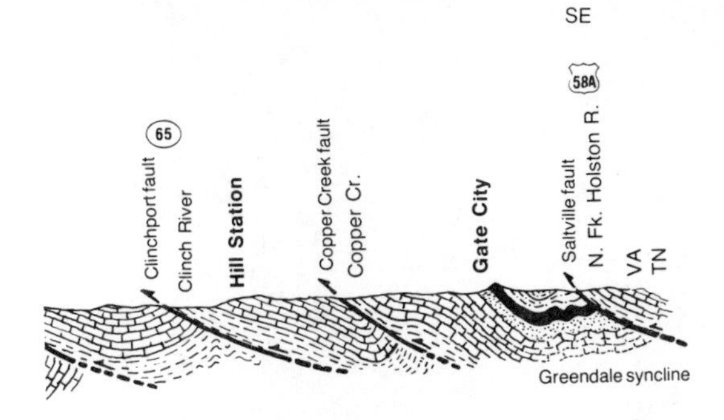

roughly parallel to US 23 from Kentucky to Tennessee.

Clinch sandstone to roadcuts in the younger Devonian Chattanooga shale.

About one-half mile south of the junction with US 58/421, the Hunter Valley thrust fault brings middle Cambrian shales and limestones onto younger Chattanooga shales. Limestones and shales in the roadcuts above the fault are much broken and faulted. South of this thrust fault the highway crosses Purchase Ridge syncline. Natural Tunnel is a cavern dissolved in Chepultepec limestone of the early Paleozoic carbonate bank in the flat core of the syncline. The tunnel is large enough to contain both Stock Creek and a railroad track. A short walk from Park Headquarters takes you to a spectacular overlook of this geologic extravaganza.

The Hunter Valley thrust fault puts middle Cambrian age limestone on top of Devonian age shale in this roadcut on US 23 near Duffield. Photo by Thomas M. Gathright, II; Courtesy of the Virginia Division of Mineral Resources.

Ripple marks in the limestone in roadcuts along the Gate City bypass indicate very shallow early Paleozoic water after sediments were deposited.

A mile south of the Natural Tunnel turn-off, the Clinchport thrust fault brings early Cambrian limestones of the Rome formation onto the early Paleozoic carbonate bank. Less than 2 miles farther along, the Copper Creek thrust fault repeats that pattern.

South of the Clinch River, the highway runs parallel to Clinch Mountain in a limestone valley cut in the early Paleozoic carbonate bank. These carbonates are predominantly dolomite, whereas formations of the same age to the east are limestone. Bedding planes exposed by the roadcut on the north side along the Gate City bypass display mudcracks and ripple marks. Evidently, the accumulating sediments were exposed at low tide.

Mud cracks in the limestones in the roadcuts on the Gate City bypass indicate that these sediments were exposed to early Paleozoic air before they were buried and lithified.

US 58, US 58A, US 19, US 23, US 421
Abingdon — Cumberland Gap
134 mi./216 km.

Between Abingdon and Hansonville, the route cuts across the trend of a series of fault slices cut in rocks of the Valley and Ridge province. Abingdon is on Cambrian rocks of the early Paleozoic carbonate bank. Just west of town, the Holston Mountain thrust fault cuts these formations, but the displacement is small, and the fault dies out a few miles to the northeast. The Pulaski thrust fault moved these Cambrian carbonates onto younger Ordovician ones. This slice extends for two miles as the road goes back into Cambrian rocks. The Saltville thrust fault moved middle Cambrian carbonates of the Honaker formation onto Mississippian rocks in the Greendale syncline.

Around Greendale the topography looks almost like that on the Appalachian Plateaus. In fact, the strip of land from just north of Greendale almost to the North Carolina state line might be considered a sliver of the plateau caught in the middle of the Valley and Ridge province. The Early Grove gas field lies in the southern portion of this sliver.

The highway crosses gently dipping layers of rock onto Devonian shales just north of the North Fork of the Holston River. The Flat Top Mountain anticline has Ordovician rocks in its core. The Clinch sandstone in its flanks stands up in high relief in Brumley and Clinch mountains. The road passes through a gap in Clinch Mountain. The rocks exposed there are Ordovician carbonate formations below the Clinch sandstone.

NW
KY
VA
Pine Mtn.
St. Paul fault
Clinch R.
Pine Mtn. fault
Middlesboro syncline

Cross section of Valley and Ridge geologic

The town of Saint Paul is at the boundary between the Appalachian Plateaus and the Valley and Ridge provinces. Here, the Saint Paul fault thrusts Cambrian folded rocks of the early Paleozoic carbonate bank up to the northwest onto flat-lying rocks of the lower Pennsylvanian sequence. Small faults appear in the roadcuts around Castlewood.

Between Appalachia and Norton, the highway skirts the boundary between the Valley and Ridge and Appalachian Plateaus provinces. Some roadcuts expose steeply tilted rock layers typical of the Valley and Ridge. Other roadcuts expose nearly horizontal rock layers characteristic of the Appalachian Plateaus. Deep roadcuts and steep cliffs in sandstone mark the scenery of the Appalachian Plateaus rocks most of the way between Saint Paul and Norton. Norton, on Pennsylvanian rocks of the Norton formation, is in coal country. The highwalls of strip mines are obvious, the roadcuts east of town expose coal seams, and low grade coal is used here and there for driveways. Most of the level ground in this area is level only because it is the remains of strip mining operations.

Coal seams in Pennsylvanian sandstones lubricate bedding plane thrust faults between St. Paul and Coeburn. Note the normal fault sloping to the right and the wedge of coal squeezed into the sandstone. Photo by Thomas M. Gathright, II; Courtesy of the Virginia Division of Mineral Resources.

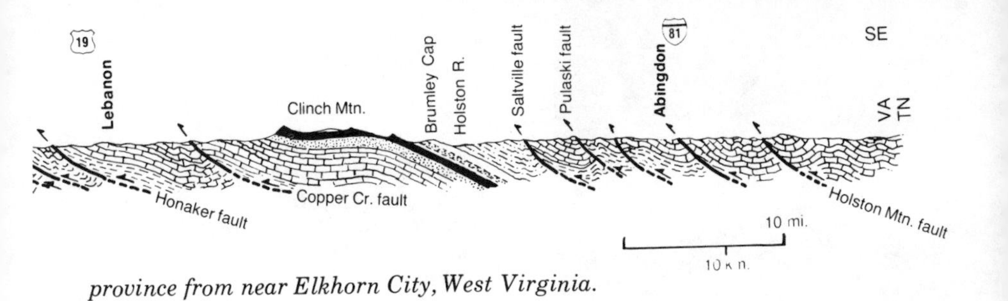

province from near Elkhorn City, West Virginia.

Big Stone Gap is in the water gap where the headwaters of the Powell River breach Big Stone Mountain, the northwest flank of the Powell Valley anticline. Layers of rock wrap around the nose of this anticline in a hairpin curve just south of Norton. The fold extends another 100 miles southwest. Many minor folds and faults within the Powell Valley anticline complicate the geologic map.

Between Big Stone Gap and Pennington Gap, the highway follows a carbonate valley between Stone Mountain and Wallen Ridge. Faults within the Powell Valley anticline brought up numerous weather-resistant Silurian sandstone slices, which form an almost chaotic array of ridges.

From Pennington Gap, US 421 goes north through the water gap in Stone Mountain where the Pennsylvanian sandstone beds stand on edge. Plateau rocks on the north side lie flat and, here again, coal mining is the major occupation. This is the southeast side of the broad, open Middlesboro syncline which extends some 10 miles to the northwest to its edge at the Pine Mountain thrust fault in Kentucky. West of Pennington Gap, the main route follows a carbonate valley for about 5 miles, then swings south across Chestnut Ridge to Jonesville. Chestnut Ridge is an anticline with a window on either side of the road. Views to the south from Jonesville are of Wallen Ridge just

Powell River carved Powell Valley in the core of the Powell Valley anticline. View is to the southwest from near the head of the valley. Photo by Thomas M. Gathright, II; Courtesy of the Virginia Division of Mineral Resources.

Along US 421 north of Pennington Gap the sandstones and shales are flat lying in the Appalachian Plateaus geologic province.

across the Powell River valley and Powell Mountain in the distance. Both are held up by Clinch sandstone. The duplication exists because the Wallen Valley thrust fault lies between the two ridges.

US 58 follows the trough of the Cedars syncline between ridges of Knox carbonates west from Jonesville. Note that the dips of the rocks are not steep. This is a typical carbonate valley and its topography much resembles that in the Shenandoah Valley, some 250 miles to the northeast. North of the highway, the white cliffs near the crest of the mountain are the Lee sandstone, long a beacon for travelers. Watch near Rose Hill for a small fault where sandstone to the east is thrust over that to the west. Carbonate rocks of the Knox group in the roadcuts show both bedding planes and solution cavities.

Between Gibson Station and Cumberland Gap, US 58 climbs Cumberland Mountain. Rocks along the way include Ordovician and Silurian sedimentary formations. The Pennsylvanian Lee sandstone caps the sequence of strata on the north side of the gap. Cumberland Gap is the pass made famous by Daniel Boone who came this way in 1775. Countless settlers followed his Wilderness Road across this pass on their way to settle the wilderness of Kentucky.

Cumberland Mountain is the boundary between the Appalachian Plateaus province to the northwest and the Valley and Ridge geologic province to the southeast. It is the northwest flank of the Powell Valley anticline and the southeast flank of the Middlesboro syncline. The ridge is held up by the Lee sandstone, with younger rocks northwest and older rocks southeast of the crest.

GEOLOGICAL JARGON

Allochthon: A fault bounded mass of rock which has been transported by tectonic processes from its site of origin to its present location. The Smith River allochthon, for example, was thrust over the Sauratown Mountains to its present location next to the Blue Ridge.

Alluvium: Unconsolidated clay, silt, sand, or gravel that has been deposited by a stream or river.

Amphibole: A group of silicate minerals, including hornblende, that is rich in iron and magnesium; one of the mafic minerals.

Amphibolite: A metamorphic rock composed predominantly of an amphibole mineral.

Anorthosite: An igneous intrusive rock composed of more than 90 percent plagioclase.

Anticline: An arching rock fold that is closed at the top and open at the bottom. The oldest formation occurs in the center of an anticline.

Aquifer: A rock formation capable of storing and transmitting ground water.

Asbestos: Any mineral that grows as a fiber.

Asthenosphere: The plastic layer beneath the lithosphere where seismic waves are attenuated and upon which the lithosphere floats.

Asymmetrical fold: An anticline or syncline wth limbs having unequal slope angles. Appalachian folds are generally steeper on their northwest limbs.

Aureole: A zone around an igneous intrusion which has been chemically or physically affected by that intrusion.

Axis: An imaginary line along each bed where there is maximum of curvature in an anticline or syncline. The axial plane of a fold connects all the axes.

Bar: Sand or other sediment in the form of a ridge along or across a stream or bay.

Barrier island: A sand ridge built by wave action parallel to a coast line. The islands off the Eastern Shore are barrier islands.

Basement: A complex of igneous and metamorphic rocks that underlies the sedimentary rocks of a region. The basement under the Coastal Plain is Paleozoic and that under the Valley and Ridge is Precambrian in age.

Basin: Either a rock structure where all the strata dip toward the center or a rift valley that has filled with sediment.

Bedding: Layers of sedimentary rocks.

Bedding-plane fault: A thrust fault which follows bedding planes, generally in a weak rock. Bedding-plane faults generally follow shale, salt, or coal beds.

Breccia: A rock composed of angular rock fragments. The Max Meadows breccia, for example, formed during substantial thrust fault movement south of Roanoke.

Carbonate bank: A shallow or intertidal continental margin maintained by carbonate secreting organisms.

Carbonate rock: Collective term including limestone and dolomite.

Carolina bay: A raised ring of sand, ranging from a few yards to more than a mile in diameter, surrounding a low area composed of clay and silt on the Coastal Plain.

Cavern: Void space in carbonate rock created by solution. The Shenandoah and other carbonate valleys overlie rock riddled by caverns.

Charnockite: A granitic rock with pyroxene as its dark mineral.

Chert: A fine-grained silica rock formed by precipitation from water.

Chlorite: A green layered silicate mineral resembling mica. Generally forms during metamorphism at a relatively low temperature.

Clastic rock: A sedimentary rock composed of fragments (clasts) of minerals or rocks that were formed elsewhere and transported to their site of deposition by wind or running water. Sandstones, siltstones, and shales are included.

Clay: A very fine-grained silicate mineral formed by the chemical alteration of pre-existing silicate minerals.

Cleavage: The tendency of a rock or a mineral to break along specific planes. Slate has rock cleavage and mica has mineral cleavage.

Columnar jointing: Vertical cleavage in an igneous rock, formed during cooling, that tends to produce generally six-sided columns.

Competence: The property of a rock to deform without breaking or flowing. Limestone and sandstone are generally more competent than shale.

Conglomerate: A sedimentary rock containing granules, pebbles, cobbles, or boulders.

Continental shelf: A gently sloping, seaward continuation of a coastal plain. Continental shelf becomes coastal plain when sea level falls.

Core: The central part of the Earth thought by geologists to be composed predominantly of iron.

Country rock: Rock present before igneous intrusion or mineral deposition by hydrothermal solution.

Cross-bedding: A sedimentary bed, commonly sandstone, which contains within it inclined layers. The dunes along the Atlantic beaches are cross-bedded.

Cross-section: A drawing of what a section of the Earth would look like if a trench were cut into it.

Crust: The outermost layer of the lithosphere. Continental crust is mostly granitic whereas oceanic crust is mostly basaltic.

Décollement: A slip surface of a bedding-plane fault. Literally, a décollement is an unsticking of the rocks.

Delta: A mass of sediment deposited at the mouth of a river.

Deposit: An accumulation of sediment or of ore minerals.

Diabase: An igneous rock similar to basalt except that the mineral grains are larger.

Dike: A tabular body of igneous rock that cuts across the structure of the rock it intrudes. Dikes form as fractures fill with magma.

Dip: The inclination of bedding or any other planar rock feature as measured from the horizontal.

Divide: A line separating two drainage basins. The Blue Ridge south of Roanoke, for example, is the divide between Atlantic and Gulf of Mexico drainage.

Dolomite: A mineral composed of calcium and magnesium carbonate or a rock composed of the mineral dolomite.

Dome: A geologic structure in which the rock layers dip away from a center in all directions, opposite of a basin.

Drowned valley: A valley flooded by a rise in sea level. All the tidal estuaries of the Coastal Plain are drowned valleys.

Dune: A mound of sand shaped by wind.

Elevation: The height of a point above mean sea level.

Epidote: A green silicate mineral that is common in the Catoctin greenstone.

Epoch: A subdivision of a period of geologic time.

Era: A division of geologic time including several periods. The eras represented in Virginia rocks are the Proterozoic, Paleozoic, Mesozoic, and Cenozoic.

Erosion: All processes that transport loose material downhill, downstream, or downwind.

Estuary: A long body of tidal water, salty at the open end and fresh at the upstream end. Virginia rivers east of the Fall Line are estuaries resulting from drowning of river valleys.

Eustatic change: Rise or fall of sea level world wide.

Evaporite: Salt and other minerals precipitated through evaporation of water. Saltville is the location of a buried evaporite deposit that is worked commercially.

Extrusive rock: Igneous rock solidified from lava or ash erupted from a volcano. Rhyolite and basalt are extrusive igneous rocks.

Facies: Sedimentary rock characteristics indicative of a particular depositional environment. Metamorphic rock characteristics indicative of a particular pressure and temperature of metamorphism.

Fault: A fracture or fracture zone in the crust where displacement has taken place.

Fault zone: The surface that best represents the location of a fault.

Feeder: A dike that fed magma to the surface.

Feldspar: The group of silicates that makes up the bulk of the crust of the Earth.

Felsic mineral: Light-colored minerals related to feldspar. Felsic rocks are those rich in felsic minerals and poor in mafic minerals.

Fissility: The tendency of a shale to break into thin flakes.

Floodplain: The area of a river that is under water during flooding.

Fold: The shape of warped rock strata including synclines and anticlines.

Foliation: The layering of leaf-shaped minerals such as micas in a metamorphic rock.

Footwall: The block of rock beneath a fault or fault zone. Miners working in a fault zone walk on the footwall.

Formation: The basic unit of rock designation. It is the fundamental [illegible] geologist plots on a map.

Fossil: Any remains of once living matter or its imprint preserved in a rock.

Gabbro: An intrusive igneous rock composed primarily of visible grains of plagioclase feldspar and pyroxene.

Garnet: A group of silicate minerals.

Geochronology: The science of determining rock ages by means of their radioactive constituents.

Geomagnetic field: The field, originating in the Earth's core, that aligns a compass needle with the magnetic north pole.

Gneiss: A metamorphic rock composed predominantly of visible feldspar grains and showing color banding or alignment of mineral grains.

Graben: A fault-bounded trough in the Earth. The Triassic basins of the Piedmont are grabens.

Granite: An igneous rock composed predominantly of visible grains of feldspar and quartz.

Granulite: A gneiss containing garnet or pyroxene or both.

Greenschist: A metamorphic rock composed largely of chlorite, a green mineral.

Greenstone: A metamorphosed basalt composed predominantly of plagioclase, green amphibole, and epidote.

Groundwater: Water contained in fractures and pores of a rock.

Half graben: A fault-bounded trough in the Earth where one side has dropped substantially more than the other. Many Triassic basins of the Piedmont are half grabens.

Hanging wall: That block of rock above a fault or fault zone. Miners working in a fault zone have the hanging wall above their heads.

Hematite: A red iron oxide mineral.

Hydrothermal deposit: Minerals deposited from hot water solutions.

Igneous rock: A rock formed by crystallization from a magma within the crust, or from a lava on the surface.

Ilmenite: An iron-titanium oxide.

Incompetence: The capacity of a rock to deform easily during folding or faulting.

Intrusive rock: An igneous rock that forced its way into a host rock and crystallized within the Earth's crust.

Inverted topography: A landscape in which streams follow former divides.

Island arc: A chain of volcanic islands.

Joint: A fracture in a rock.

Karst topography: A landscape characterized by sinkholes, sinking streams, and caves formed through solution of underlying limestone or dolomite.

Downhill slippage of a mass or soil and rock.

has reached the Earth's surface.

tituents from soil, rock, landfill, and so forth.

d.

edominantly of the mineral calcite.

Lineament: Any linear feature of the Earth's surface.

Lineation: Any alignment of mineral needles or rods in a metamorphic rock.

Lithosphere: The outer rocky shell of the Earth including the oceanic and continental crust and the mantle down to the asthenosphere. The lithosphere is composed of tectonic plates.

Longshore drift: Lateral movement of sand along a coast driven by breaking waves and backwash.

Mafic mineral: A dark-colored silicate mineral rich in magnesium and iron and including pyroxene, amphibole, and biotite mica. A mafic rock contains large amounts of mafic minerals.

Magma: Molten rock within the Earth. Magma that reaches the surface of the Earth is lava.

Mantle: That part of the Earth between the crust and the core including the lower lithosphere and the asthenosphere.

Marble: A metamorphic rock composed of calcite or dolomite.

Matrix: Fine-grained minerals surrounding coarse crystals or fragments.

Mean sea level: Defined elevation between high and low tides from which elevations and depths are measured.

Metamorphic grade: Relative intensity from low to high of metamorphic pressures and temperatures.

Metamorphism: Mineral and texture changes in a rock due to heat and pressure from burial in the crust or from a nearby igneous intrusion.

Mica: Silicate mineral that cleaves into thin sheets. The chief micas are black biotite and colorless muscovite.

Mineralization: The introduction of new minerals into a host rock, generally by water solution.

Mountain range: Linear band of folded, faulted, and partially metamorphosed rock.

Mylonite: Lithified fine-grained material produced in a fault zone.

Normal fault: A fault in which the block above the fracture drops with respect to that below.

Obduction: Addition of a part of a subducting tectonic plate onto the overriding plate.

Ophiolite: A mix of oceanic igneous and sedimentary rock that has been obducted onto continental crust.

Orogeny: A tectonic process which results in formation of mountain belts. Orogenies affecting Virginia rocks include the Grenville (1000 million years), Taconic (450 million years), Acadian (350 million years), and Alleghanian (250 million years).

Outcrop: Exposed bedrock.

Overthrust belt: A zone in which older rocks were shoved over younger rocks.

Overturned fold: An anticline or syncline in which one limb has been turned past vertical.

Paleomagnetism: A record of past geomagnetic fields impressed on magnetic minerals in a rock.

Paleontology: The study of fossils.

Pangaea: The single continent that coalesced at the end of the Paleozoic era.

Panthalassa: The single world ocean that surrounded Pangaea.

Pegmatite: Igneous rock characterized by large crystals. Most pegmatites are close to granite in composition.

Period: A unit of geologic time.

Permeability: The capacity of a rock to transmit water, oil, or natural gas.

Petrology: The study of rocks.

Phyllite: A metamorphic rock composed predominantly of fine-grained mica.

Piracy: A geologic process in which one stream captures the headwaters of another by means of erosion.

Placer: An accumulation of economically valuable minerals in beach or stream deposits.

Plagioclase: A calcium to sodium feldspar.

Plastic deformation: Rock deformation such as folding without breakage.

Plate: A segment of the Earth's lithosphere that moves about the globe independently of other plates. Lithospheric plates move at about the same rate as fingernails grow. In overthrust belts, a hanging-wall block may be referred to as a plate.

Plate tectonics: A modern analysis of ocean formation and mountain building in terms of interactions between lithospheric plates.

Pluton: A body of igneous rock that crystallized within the crust of the Earth.

Porosity: Open space in a rock that may contain air, water, oil, or natural gas.

Pothole: A hole in the bedrock of a stream caused by abrasion by cobbles or boulders agitated by strong currents.

Precambrian time: All geologic time before 600 million years ago.

Proterozoic Era: That period of geologic time extending from 2500 million years until the Phanerozoic Era began some 600 million years ago.

Pyrite: Yellow, metallic iron sulfide.

Pyroclastic rock: A volcanic rock composed of fragments of various sizes resulting from explosive vulcanism.

Pyroxene: A group of silicate minerals rich in iron and magnesium which occurs in dark igneous rocks and in high-pressure metamorphic rocks.

Quartz: A mineral composed of pure silica.

Quartzite: A sedimentary or metamorphic rock composed of quartz grains cemented together by silica.

Radiometric age: The age of a rock as determined by analysis of radioactive elements and their decay products.

Recharge: The infiltration of rain water into pore space in soil and rock.

Recrystallization: The growth of new minerals in a rock as a result of changes in pressure, temperature, or composition of pore fluids.

Reef: A linear or circular ridge of coral and other lime secreting organisms.

Regional metamorphism: Recrystallization of originally igneous or sedimentary rocks due to burial in the crust over a wide area.

Relief: Differences in elevation on a local or a regional basis.

Retrograde metamorphism: Mineral change in a rock brought about by falling pressure and temperature.

Reverse fault: A steep fault in which the block above the fracture rose with respect to that below.

Rhyolite: A volcanic rock composed predominantly of quartz and feldspar.

Rift: A trough formed by extension of the crust.

Ripple: A small ridge of sand or silt.

Rock: A solid aggregate of one or more minerals making up a unit of sufficient size for geologists to plot on a map.

Rootless anticline: An anticline which is terminated at depth by a fault or fault zone. Appalachian anticlines are all rootless.

Runoff: Rainwater running off the land surface as opposed to that which soaks into pore space in soil and rock.

Rutile: A red titanim oxide mineral.

Sandstone: A rock composed of sand grains cemented together into a cohesive unit.

Saprolite: Clay, sand, and iron oxide resulting from complete degradation of bedrock. Piedmont bedrock is altered to saprolite in some places to depths exceeding 100 feet.

Scarp: A rise between two land surfaces of different elevations. Scarps range in size from a few feet on the Coastal Plain to more than a thousand feet for the southern Blue Ridge scarp.

Schist: A metamorphic rock containing sufficient platy minerals such as mica to give it a pronounced foliation and rock cleavage.

Sediment: Natural materials deposited by wind or running water on the surface of the Earth.

Serpentine: A magnesium silicate mineral group indicative of low-grade metamorphism.

Shale: A sedimentary rock composed of clay and silt grains which splits along narrowly spaced bedding planes.

Shield: A tectonically stale continental region underlain by crystalline rocks.

Sill: A tabular intrusive igneous rock that is parallel to its host bedding planes.

Sinkhole: A surface depression resulting from rock solution and cavern collapse.

Slate: A fine-grained metamorphic rock that splits cleanly into thin slabs.

Slump: Downslope collapse of soil and rock.

Soapstone: A massive metamorphic rock composed primarily of talc and serpentine. Most soapstones result from low-grade metamorphism of oceanic igneous rocks.

Soil: Sand, silt, and clay on the Earth's surface resulting from weathering of bedrock.

Spit: A ridge of sand created by longshore drift of beach materials.

Stalactite: A cone of precipitated mineral, generally calcite, hanging from the ceiling of a cavern.

Stalagmite: A cone of precipitated mineral rising from the floor of a cavern.

Strata: Layers of sedimentary rock.

Strike: The angle between north and a horizontal line on any planar rock feature such as bedding or jointing.

Strike-slip fault: A steeply dipping fault that moves horizontally.

Subduction: In plate tectonics, when one lithospheric plate overrides another, the overridden plate is subducted into the Earth's mantle. The Paleozoic Iapetus Ocean was closed by subduction.

Subsidence: The sinking of a broad area of Earth's crust without appreciable deformation of rock strata.

Superposed stream: A stream that eroded its channel downward onto hidden geologic structures.

Symmetrical fold: A rock fold with limbs having the same dip angle.

Syncline: A rock fold that is closed at the bottom and open at the top. The youngest rocks are at the center of a syncline.

Synclinorium: A broad fold in the crust containing many smaller synclines and anticlines.

Talc: A platy magnesium silicate mineral. Pure talc is crushed, perfumed, and sold as talcum powder.

Tectonics: Study of crustal rock movement and deformation on a large scale. Plate tectonics is the study of the interactions of lithospheric plates with an objective of determining Earth history.

Terrain: Character of a land surface.

Terrane: A fault-bounded volume of the Earth's crust. Adjacent terranes have substantially different geologic histories and were juxtaposed by subsequent tectonic transport.

Texture: Of a rock, the characteristics of grain or crystal size, shape, and orientation.

Thrust fault: A fault in which the block above the fracture is shoved over that below at a low angle.

Tidal current: Water movement in bays and estuaries resulting from rise and fall of ocean tides.

Tidal flat: A flat, muddy area exposed at low tide but submerged at high tide.

Topography: The shape of the Earth's surface.

Transgression: Submergence of an area due to rise in sea level or subsidence of the land surface.

Travertine: Limestone deposited in caverns or around hot springs.

Ultramafic rock: An igneous rock with little feldspar and large amounts of mafic minerals.

Unconformity: A buried rock surface that represents an interval of nondeposition or erosion of underlying rock formations. Coastal Plain sediments rest unconformably on crystalline basement rocks.

Underfit stream: A stream that is too small to account for the size of its valley and floodplain.

Uplift: Increase in elevation over a broad region of the crust.

Vein: A mineral deposit along a joint or fault in the host rock.

Vesicle: A bubble in a volcanic rock. Most vesicles in Catoctin greenstones are filled with calcite, quartz, or other minerals.

Volcanic ash: Small fragments of lava blasted from an explosive volcano.

Volcanic plug: A cylinder of igneous rock trapped in the throat of an ancient volcano.

Water table: The surface in the ground dividing water-saturated rock below from unsaturated rock above. The level of water in a well indicates the elevation of the water table.

Weathering: All the processes that lead to decomposition of bedrock and formation of talus, soil, and saprolite.

Xenolith: Any chunk of pre-existing rock trapped in a crystallizing igneous rock.

Zeolite: A class of silicate minerals that give up water upon heating.

ADDITIONAL READING

Catlin, D. T., 1984. *A Naturalist's Blue Ridge Parkway*. Knoxville: University of Tennessee Press, 208 p. A guidebook to the rocks, flora, and fauna of the Blue Ridge.

Dietrich, R. V., 1970. *Geology and Virginia*. Charlottesville: University Press of Virginia, 213 p. A textbook emphasizing Virginia geology but without reference to global tectonics.

Dietrich, R. V., 1970. *Minerals of Virginia*. Research Division Bulletin 47, Virginia Polytechnic Institute, Blacksburg, 325 p. An alphabetical listing of the minerals found in Virginia with mineral descriptions and locations.

Dietrich, R. V., and B. J. Skinner, 1979. *Rocks and Rock Minerals*. Englewood Cliffs, New Jersey: Prentice-Hall, 319 p. An introduction to the classification and identification in hand specimen of common rocks and minerals.

Eicher, D. L., A. L. McAlester, and M. L. Rottman, 1984. *The History of the Earth's Crust*. Englewood Cliffs: Prentice-Hall, 198 p. Global tectonics applied to Earth history.

Fairbridge, R. W., 1975. *Encyclopedia of World Regional Geology*, Part 1, Western Hemisphere (Including Antarctica and Australia). Encyclopedia of Earth Sciences Series volume VIII. Stroudsburg: Hutchinson Ross, 736 p. Places Virginia within the continental geologic framework.

Gathright, T. M., II, 1976. *Geology of the Shenandoah National Park, Virginia*. Bulletin 86, Virginia Division of Mineral Resources, Charlottesville, 93 p., 3 maps.

Publications, Virginia Division of Mineral Resources, Charlottesville. Technical reports and geologic maps at irregular intervals.

Renfro, H. B., and D. E. Feray, 1970. *Geological Highway Map, Mid-Atlantic Region*. Tulsa: American Association of Petroleum Geologists, 1 sheet. Color geologic map of Virginia and surrounding states.

Virginia Minerals. Division of Mineral Resources, Charlottesville. Quarterly publication on a variety of geologic topics.

Wyllie, P. J., 1976. *The Way the Earth Works: An Introduction to the New Global Geology and its Revolutionary Development*. New York: Wiley, 296 p. Textbook emphasizing global tectonics.

A current list of publications and maps may be requested from Virginia Division of Mineral Resources, Box 3667, Charlottesville, Virginia 22903. Their public sales office is located on McCormick Road on the University of Virginia campus.

Index